Strategies for Sustainability

For further volumes:
http://www.springer.com/series/8584

Strategies for Sustainability

The series, will focus on "implementation strategies and responses" to environmental problems – at the local, national, and global levels.

Our objective is to encourage policy proposals and prescriptive thinking on topics such as: the management of sustainability (i.e. environment-development trade-offs), pollution prevention, clean technologies, multilateral treaty-making, harmonization of environmental standards, the role of scientific analysis in decision-making, the implementation of public-private partnerships for resource management, regulatory enforcement, and approaches to meeting inter-generational obligations regarding the management of common resources. We will favor trans-disciplinary perspectives and analyses grounded in careful, comparative studies of practice, demonstrations, or policy reforms. We will not be interested in further documentation of problems, prescriptive pieces that are not grounded in practice, or environmental studies. Philosophically, we will adopt an open-minded pragmatism – "show us what works and why" – rather than a particular bias toward a theory of the liberal state (i.e. "command-and-control") or a theory of markets.

We invite Authors to submit manuscripts that:

– Prescribe how to do better at incorporating concerns about sustainability into public policy and private action.
– Document what has and has not worked in practice.
– Describe what should be tried next to promote greater sustainability in natural resource management, energy production, housing design and development, industrial reorganization, infrastructure planning, land use, and business strategy.
– Develop implementation strategies and examine the effectiveness of specific sustainability strategies. – Focus on trans-disciplinary analyses grounded in careful, comparative studies of practice or policy reform. – Provide an approach "...to meeting the needs of the present without compromising the ability of future generations to meet their own needs," and do this in a way that balances the goal of economic development with due consideration for environmental protection, social progress, and individual rights.

Submit your Proposal!

Authors interested in submitting a book proposal under these series should contact the Publishing Editor for a Book Proposal Form:

Fritz Schmuhl, Senior Editor: Environmental Economics, Management, and Policy, Springer Dordrecht, Van Godewijckstraat 30, P.O. Box 17, 3300 AA Dordrecht, The Netherlands. Phone: +31 78 657-6281, e-mail: Fritz.Schmuhl@springer.com

Peter A. Wilderer • Martin Grambow
Editors

Global Stability through Decentralization?

In Search for the Right Balance between Central and Decentral Solutions

Springer

Editors
Peter A. Wilderer
Institute for Earth System Preservation (IESP)
Technical University of Munich
Garching, Germany

Martin Grambow
Bayerisches Staatsministerium für Umwelt und Verbraucherschutz
Munich, Germany

Strategies for Sustainability
ISBN 978-3-319-24356-6 ISBN 978-3-319-24358-0 (eBook)
DOI 10.1007/978-3-319-24358-0

Library of Congress Control Number: 2015958666

Springer Cham Heidelberg New York Dordrecht London

Printed on acid-free paper

Springer International Publishing AG Switzerland is part of Springer Science+Business Media (www.springer.com)

Preface

Inspired by the discussion about application of technology designed to counteract global warming and the resulting climate change, often referred to as "geoengineering", the Institute of Advanced Studies on Sustainability of the European Academy of Sciences and Arts organised a workshop entitled "Earth System Engineering—The Art of Dealing Wisely with the Planet Earth" (wbk1). The workshop was held in 2008 at the premises of the Hanns Seidel Foundation at Wildbad-Kreuth, Germany. The Institute for Advanced Study of the Technical University of Munich (TUM-IAS) generously supported that workshop. On advice of the participants, the International Expert Group on the Preservation of the well-functioning of the Earth System (IESP) was established. Since then, IESP acts as an open access information network composed of scientists, entrepreneurs, administrators and politicians. Over a period of 6 years (2008–2015), the Bavarian Ministry of Environment provided financial support administered by TUM-IAS.

The main purpose of IESP was and is to organise seminars and workshops on topics of general interest to private and public decision makers. IESP considers itself as a Think Tank on issues specifically focused on the interconnectedness of the three major domains of the Earth system: Ecology, Economy and Society. Sustainable development of this eco-social triad requires knowledge to be provided by science. Leaders in society, economy and political institutions are invited to draw decisions on the basis of scientifically consolidated knowledge with the ultimate aim to preserve life-enabling conditions on Earth.

The workshop mentioned above (wbk1) led to the conclusion that science is by far not yet in the position to engineer our planet. Because of the incomplete knowledge of the complex, mostly non-linear nature of the Earth system application of intended measures to manipulate the Earth system bears the risk of driving the Earth system to a point of no return, often called tipping point, leading to disasters and eventually to collapse. In short, the meeting revealed two important insights:

1. The impact of any attempt to counteract global warming and the resulting climate change through geoengineering methods would bear more risks than opportunities.
2. The paradox that technical advances—as beneficial they are—have become a threat to our civilisation. Climate change, water scarcity, disturbance of the bio-system and social tensions are just some of the examples of the global problems we are nowadays facing.
3. Most participants came to the conclusion that only extended sustainability strategies could lead the way out of this trap.

The recommendations resolved by the participants of the workshops are presented in the appendix to this book.

The first follow-up workshop held at Wildbad-Kreuth in 2012 (wbk2) was focused on the resilience theory. Here, the question was whether strengthening the resilience of either of the subsystems (economy, ecology, society) of the eco-social triad (Adams 2006) is a promising step forward to sustainability.

Recent history has shown that the process towards sustainability has achieved only tangible results. This is mostly due to sociopolitical reasons, but also because of the lack of a readily understandable definition of the term "sustainability". The discussions during the workshop led to the common understanding of resilience being a precondition of sustainable development.

Resilience is to be understood as a dynamic state, a moving target that has to be permanently approached through an iterative process of continuous adjustment to the existing environmental conditions and to temporary disturbances. Sustainable development is based on strategies that will minimise the likelihood of the eco-social triad to exceed any tipping point, when experiencing severe stress or when exposed to major disturbances (Bloesch et al. 2015).

The participants of the wbk2-workshop came to the conclusion that keeping the Earth system in balance requires readiness of all actors to take in consideration and proactively respond to the ongoing changes of ambient conditions on the local as well as on the global scale, be it climatic, economic, political or societal conditions. As history teaches us, that attempts to conserve the *status quo* sooner or later destabilise the system of concern making it vulnerable and prone to collapse.

The contributions made in response to the third Wildbad-Kreuth workshop (wbk3) are presented in this book. The organisers of this workshop expressed doubts that keeping the Earth system in balance could be achieved by globalised decision-making, for instance on the level of the United Nations. Since most of the global changes originate on the local scale, and since the local climatic, economic, political or societal conditions vary significantly across the planet, it was hypothesised that decentralisation of decision-making is the more promising way to keep the Earth system in the state of resilience. In this book, authors from a wide variety of disciplines present their views on the controversy of centralisation and decentralisation in search for the most promising way to keep the Earth system in balance.

The recommendations resolved by the participants are presented in the form of a declaration (Sect. 7.1).

Garching, Germany — Peter A. Wilderer
Munich, Germany — Martin Grambow

References

Adams WM (2006) The future of sustainability: Re-thinking environment and development in the twenty-first century. Report of the IUCN-renowned thinkers meeting 29–31 Jan 2006, 19 pp

Bloesch J, von Hauff M, Irvine RL, Mainzer K, Mohan SV, Renn O; Risse V, Schroeder ED, SongY, Takeuchi K, Wilderer PA (2015) Sustainable development goals integrated in the concept of resilience. Problemy Ecorozwoju—Problems of Sustainable Development, pp 10–11 (in press)

Contents

1 Regulation and Control Processes in Ecosystems 1
Wolfgang Haber and Jürg Bloesch
1.1 Ecosystem Function 1
1.1.1 Key Messages 1
1.1.2 What Is an Ecosystem? 1
1.1.3 Ecosystem Structure and Organization 2
1.1.4 Ecosystem Functioning and Regulations: Without Central Control 5
1.1.5 The Arrival of Humans in Nature 6
1.1.6 Formation of the Human Environment 7
1.1.7 Rural and Urban Living: The Anthroposphere 7
1.1.8 Natural Ecosystems Compared to Human-Made Systems . . . 9
1.1.9 Conclusion 9
1.2 Globalization and Balance: The Ecosystem Dimension 11
1.2.1 Key Messages 11
1.2.2 Introduction 11
1.2.3 The Role and Function of Natural Ecosystems 11
1.2.4 Ecosystem Service and Human Use: Artificial Human Ecosystems 12
1.2.5 Natural and Human-Made-Growth: The Impact of Economy 14
1.2.6 The Philosophy Behind: Ethics, Religion and Education Matter 16
1.2.7 Balance Between Centralization and Decentralization in Technical Systems 17
References 18

2 Control and Management of Man-Made Systems 21
Patrick Dewilde, Ulrich Drost, Martin Grambow, Bernhard Schätz, and Martin Korndörfer
2.1 Centralized Versus Decentralized Systems: A Critical Appraisal . . . 21
2.1.1 Key Messages 21
2.1.2 Example of Systems 22
2.1.3 Properties of Local Versus Global 23
2.1.4 The Design Issue 25
2.1.5 Reconciling Local with Global 26
2.1.6 Bringing Intelligence into the Control Loop 27
2.1.7 Optimizing System Behavior 29
2.1.8 Designing (Distributed) Systems 32
2.1.9 A Potpourri of Illustrative Examples 35
2.1.10 Ethics 35
2.1.11 Desirable and Actual Ethics 38
2.1.12 The Smart Grid as an Example 38
2.1.13 Final Remarks 43
2.2 The Principles of Subsidiarity and Internality as Ordering Principles of Decentralized Structures, Illustrated by the Example of Water Management 43
2.2.1 Key Messages 43
2.2.2 Subsidiarity and Internality within the Scope of Future Development of Decentralized Structures 44
2.2.3 General Legal Guidelines for Water Management Considering the Principle of Subsidiarity 45
2.2.4 Discussion Regarding the Forms and Further Development of the Principle of Subsidiarity in the Legal Requirements for Water Management 51
2.2.5 The Principle of Internality 53
2.2.6 Subsidiarity and Internality Within the Scope of Future Development of Decentralized Structures 54
2.3 Smart Energy Systems: Decentralization, Virtualization and Hierarchization 55
2.3.1 Key Messages 55
2.3.2 Renewables and Reliability: A Contradiction? 56
2.3.3 Smart Distribution: Virtualization, Decentralization, Hierarchization 58
2.3.4 Smart Grid: Resilient Balance of Demand and Supply . . . 61
2.3.5 Conclusion 63
2.4 The Importance of Negative Feedback 64
2.4.1 Key Messages 64
2.4.2 Introduction 64
2.4.3 Capital and the Neoliberal Monetary System in the Context of the Adaptive Cycle 65

2.4.4 Anthropogenic Systems Emulating Natural Systems 68
2.4.5 Conclusion 70
References 71

3 **Governance of Societies** 75
Christian Werthmann, Ortwin Renn, Anastassia Makarieva, Victor Gorshkov, Peter A. Wilderer, Verena Risse, Carolin Böker, and Bettina Haas
3.1 Skipping Centralization 75
3.1.1 Key Messages 76
3.1.2 Where and How Urbanization Will Occur 76
3.1.3 The Case of São Paulo 78
3.1.4 Skipping Centralization 80
3.2 The Balance Between Efficiency, Effectiveness, Resilience and Cohesion 80
3.2.1 Key Messages 81
3.2.2 Basic Considerations 81
3.2.3 Conclusion 83
3.3 What Can We Learn from Natural Ecosystems to Avoid a Civilization Breakdown? 83
3.3.1 Key Messages 84
3.3.2 Introduction: Does Civilization Progress Have a Direction? 84
3.3.3 Informational Precipice Between the Biosphere and Civilization 86
3.3.4 Biotic Regulation of the Environment 88
3.3.5 Principles of Order Maintenance in Life: Decentralization Is Key 91
3.3.6 Information Losses During Evolution of Large Animals 93
3.3.7 Why Do Large Animals Exist? 94
3.3.8 Ecological Rights of Animals and Man 95
3.3.9 Principle Differences Between Animals and Man 97
3.3.10 Scientific and Technological Progress and Ecological Human Needs 99
3.3.11 Conclusions 102
3.4 Nature and Human Nature: Ethical Concerns Should Not Be Disregarded in the Process of Decentralization 104
3.4.1 Key Messages 104
3.4.2 Introduction 105
3.4.3 Global Problems: Local Effects 105
3.4.4 Taking into Account Ethical Concerns 106
3.4.5 Conclusion 107

3.5 Transformation Towards a Resilient and Humane Environment and Culture: What Needs to Be Done? . . . 107
3.5.1 Key Messages . . . 107
3.5.2 Introduction . . . 107
3.5.3 How to Define "Humane Living Conditions"? . . . 109
3.5.4 Basic Points of Transformation . . . 110
3.5.5 Conclusion . . . 111
References . . . 112

4 Towards Sustainable Economies . . . 117
Michael von Hauff, Franz-Theo Gottwald, Katharina Stöckl, Jelena Kurz, and Carolin Böker
4.1 Establishment of a Sustainable Economy Requires a Proper Balance Between Centralized and Dezentralized Structures . . . 117
4.1.1 Key Messages . . . 117
4.1.2 Introduction . . . 118
4.1.3 Requirements for a Sustainable Economy . . . 119
4.1.4 The Concept of a Sustainable Industrial Park . . . 122
4.1.5 Requirements for a Sustainable Resource Strategy . . . 123
4.1.6 Conclusions . . . 125
4.2 Is Bioeconomy the Road to Decentralization? . . . 125
4.2.1 Key Messages . . . 126
4.2.2 Bioeconomy: The Economization of Life . . . 126
4.2.3 Bioeconomy and Agricultural Production . . . 127
4.2.4 Aggressive Lobbying: Disproportional Funding . . . 128
4.3 Sustainable Economy in a Globalized World: Models and Solutions to Induce Sustainable Development . . . 130
4.3.1 Key Messages . . . 130
4.3.2 Introduction . . . 130
4.3.3 Current Issues . . . 131
4.3.4 Models and Solutions for a Sustainable Economic Development . . . 132
4.4 Thoughts from a Non-economical Point of View . . . 135
References . . . 136

5 Sanitary Engineering: Central or Decentral Solutions? . . . 139
Peter A. Wilderer, Martin Grambow, Asher Brenner, and Werner P. Bauer
5.1 Urban Water Management Requires Both Central and Decentral Solutions . . . 139
5.1.1 Key Messages . . . 139
5.1.2 Introduction . . . 140
5.1.3 Water Re-Use, an Option to the Overcoming of Water Shortages in Urban Areas . . . 141

5.1.4 Livable Cities Need a Livable Hinterland 145
5.1.5 Technical and Political Challenges 146
5.1.6 Concluding Remarks 147
5.1.7 Acknowledgement 149
5.2 Decentralization of Wastewater Treatment and Reuse 149
5.2.1 Key Messages 149
5.2.2 Introduction 150
5.2.3 Is Nature Undefeated? 150
5.2.4 Decentralized Wastewater Management 150
5.2.5 Wastewater Treatment Standards for Centralized and Decentralized Systems 151
5.2.6 Existence of Sewer Systems as a Key Factor in Decision . . . 151
5.2.7 Conclusion 152
5.3 Centralized and Dezentralized Structures in Solid Waste Management 152
5.3.1 Key Messages 152
5.3.2 Introduction 153
5.3.3 Landfill 155
5.3.4 Incineration 156
5.3.5 Drivers of Recycling 159
5.3.6 Drivers of Reduction 160
5.3.7 From “Reduce: Reuse—Recycling” to “Zero Waste” 161
5.3.8 Waste Management 2030+ 162
References 162

6 Communication, Mobility and Logistics 165
Klaus Mainzer, Verena Huber, and Joachim Schütter
6.1 Challenges of Complexity in the Age of Globalization and Big Data 165
6.1.1 Key Messages 165
6.1.2 Overview 165
6.1.3 Globalization and Systems Dynamics 166
6.1.4 Information and Communication Technology (ITC) and Sociotechnical Systems 172
6.1.5 Reminder of the Internet as Complex Dynamical System . . . 172
6.1.6 Smart Grids as Complex Dynamical Systems 173
6.1.7 Example: Automatic Negotiations of Virtual Agents 173
6.1.8 Example: Internet of Things 175
6.1.9 Complexity and Big Data Technology 176
6.2 Reflections 178
References 180

7 **The Wildbad-Kreuth-Declaration** 181
Peter A. Wilderer and Martin Grambow
7.1 Recommendations 181
7.2 Explanations 183

Erratum to: Global Stability through Decentralization? E1

Appendix 193
Zugspitze Declaration on the Responsibility of Humanity for the Functioning of the Earth System 193

Explanations 195

Acknowledgement 197

Index 199

List of Contributors

Werner Bauer received his "Diplomingenieur" in civil engineering at the Technical University of Munich, Germany. He worked for 2 years as a scientist at the Technical University of Munich and started in 1985 his own consulting office. In the following years, he was responsible for several visionary solutions in waste management, like the first sorting plant at an international airport, a special building concept of a composting plant and for example the greatest benchmark platform for municipalities in Germany. An applied research work about "Synergistic Effects by Thermal and Biological Waste Treatment Facilities" shows that he always tries to bridge the ideological gap between recycling and incineration. Current activities: Managing Director of ia GmbH—Knowledge Management and Engineering, Managing Director of WtERT Germany GmbH and Vice President of Global WtERT Council. Werner Bauer has approximately 80 published articles/conference papers and 3 books as an editor in the fields of waste management and knowledge management.

Jürg Bloesch is a mandatary of the International Association for Danube Research (IAD) for which he served as National Representative (1995–1998), President (1998–2004) and Editor of Danube News (2006–2012). Born in Switzerland and educated as limnologist (hydrobiologist) at ETH Zürich, he worked at Eawag, Dübendorf, from 1970 till 2005 as research scientist and consultant. Key activities focused on lakes (sedimentation), rivers (benthos, fish) and river basin management (hydromorphology). As active member of 9 national and international professional associations and NGOs, Jürg is dedicated to environmental protection and nature conservation by bridging the gap between basic and applied science.

Carolin Böker studied Biology at the Technical University of Munich. She is about to complete her Ph.D. in Natural Sciences at the Chair of Aquatic Systems Biology of the Technical University of Munich. Her research covers interactions of benthic macroinvertebrates and vertebrates in freshwater substrata with physico-chemical habitat properties and the diversity of bacteria—key players for freshwater ecosystem functioning.

Asher Brenner is currently serving as the chairman of the Environmental Engineering programme at the Ben-Gurion University of the Negev (BGU). Asher received his Ph.D. in Environmental Engineering at the Technion, Israel (1986). He joined BGU in 1988, after a post-doc at the University of Notre Dame, USA. During his tenure at BGU, he spent a sabbatical in UC Davis, USA (1995–1996), and in Monash University, Australia (2010). Asher has published more than 70 refereed journal papers and presented numerous papers at international conferences. He has also supervised more than 50 Ph.D. and M.Sc. students.

Patrick Dewilde has been a professor in electrical engineering at the Technical University of Delft for 31 years, director of the Delft Institute for Micro-electronics and Sub-micron Technology for 10 years, chairman of the Dutch Technology Foundation STW (a major Dutch research funding agency) for 8 years and, finally, director of the Institute for Advanced Study of the Technical University of Munich for 5 years. His research has focused on mathematical issues related to the design, control and operations of dynamical systems. He is an IEEE Fellow, has been elevated to the rank of Knight of the Dutch Lion and is presently an honorary professor both at the Technical University of Munich and at the Technical University of Wroclaw.

Ulrich Drost MR a. D. is a lawyer and was 2011 leader of the department of water rights in the Bavarian State ministry of environment and consumer protection. He is an author of the water right comments "The new water right in Bavaria" and "The new water right" as well as the textbook "The new water right". Besides other numerous publications, he has taken stand for water-juridical problem formulations.

Victor Gorshkov holds position of leading researcher in the Theoretical Physics Division of Petersburg Nuclear Physics Institute in Russia. Victor was educated as theoretical physicist at the University of Leningrad, Russia. Between 1958 and 1970, Victor worked in Leningrad Physical and Technical Institute and then until now in Petersburg Nuclear Physics Institute. Victor's research interests evolved from atomic physics, quantum electrodynamics and high-energy physics to theoretical investigations of the physical and biological principles of life stability. Between 1978 and 1998, Victor was lecturing in Leningrad Polytechnical Institute on ecological physics. Victor's main scientific challenge is to quantify the stabilising impact of the natural ecosystems on Earth's environment and climate.

Franz-Theo Gottwald, Director of the Schweisfurth-Foundation, Munich. Educational background: Studies in Theology (Dipl.Theol.), Indology, Social Science and Philosophy with a Ph.D. in 1982.

Management training and consulting. Research and lecturing on political ecology at Munich School of Political Science and on CSR at Humboldt University Berlin (Hon. Professorship), Institute of Resource Economy.
Member of the Herausgeberbeirat Zeitschrift für Umweltpolitik und Umweltrecht; Chair Consumer Commission, Bavarian State Ministry of the Environment and Consumer Protection.

Fields of interest: Business Ethics, sustainability along the food chain, consciousness research and future studies. Author of numerous books and articles on the future of agriculture, education, environmental and food ethics.

Martin Grambow studied civil engineering at the TUM (Dipl.-Ing in 1986). He completed his Dr.-Ing. research in 2005 (Prof. Martin Faulstich) and was appointed an honorary professor in Mai 2012.

Martin Grambow is a water administrator, living and working in Bavaria. Since 2006, he is Director General Water in the Ministry of Environment and Health in Munich. This department leads nearly 3000 experts in the fields of engineering, natural scientist and law. They together are responsible for administrating the commons "water" and"soil" in Bavaria.

As typical for a Bavarian government official after his study, he completed a special 2 years training program in politics and administration with the final qualification as "Regierungsbaumeister". After that, he was public employee in several positions in the state government as well as in the local water office in Weilheim, the ministry of the interior, as chief of the local water office in Hof and finally different positions in the Ministry of the Environment.

In 1998, he established the state project "Technology Transfer Water". In this framework, he supported with his team international projects of institutions like the World Bank, the European Union or, in particular, the Professional Training Centre (bfz) in Hof. He has professional partnerships in Central and Eastern Europe, the USA, Brazil, Central Asia and China.

He is a member of several institutions and foundations, in particular the "Institute for Advanced Studies on Sustainability"and the "International Expert Group on Earth System Preservation" (IESP) and the "Bavarian Water Foundation".

Positions

Since 2006 Director General Water in the BavStMUG

2003–2006 Head of Unit in the BavStMUG: Water in Rural Areas, Hydraulics

1996–2006 Chief of the Regional Water Office in Hof, Upper Franconia

1992–1996 Deputy—Head of Unit Water Supply Bavaria (Ministry)

1988–1992 Head of Unit district of Bad Tölz—Wolfratshausen in the Regional Water Office in Weilheim

Bettina S. Haas works at the Bavarian State Ministry of the Environment and Consumer Protection in the unit "soil protection, contaminated sites and geology". She completed her studies of geo-ecology at the University of Bayreuth and held different positions within the Bavarian water management administration during the last 15 years.

Wolfgang Haber finished his studies of biology, chemistry and geography with a Ph.D. thesis on soil biology. From 1966 to 1994, he held the chair of landscape ecology of the Technical University of Munich with research and teaching in landscape, vegetation and ecosystem science, conservation and land use ecology, listing about 460 publications on these topics. He was President both of the Ecological Society of the German-speaking countries and of the International Association of Ecology and Visiting Professor in Austria, Switzerland, Spain, Japan and China. Haber served also as scientific advisor for environmental policy and received many honorary positions and awards.

Verena Huber studied Biology at the Technical University of Munich. She started her Ph.D. in Natural Sciences at the Chair of Aquatic Systems Biology of the Technical University of Munich in the year 2013. Her research covers the interactions of the larvae of the threatened pond mussels and potential host fish species.

Martin Korndoerfer graduated in 2004 at the Technical University of Munich with a B.Sc. in biochemistry. Between 2007 and 2011, he received his postgraduate diploma and his M.Sc. in environmental sciences from the University of Canterbury, NZ, on the topic of human–environment interactions and their effects around the Ngel Nyaki Forest Reserve, Nigeria. After returning to Germany, he worked as an environmental consultant using the Cradle-to-Cradle(R) design principles and certification system. He has been a founding member and working in an honorary capacity for the hakuna matata project, a community-based organisation for sustainable development in Kenya since 2011. He joined the Cradle-to-Cradle initiative established at the Department of Architecture of the Technical University of Munich in 2013. Since 2015, he serves the County of Munich as a manager for climate protection affairs.

Jelena Kurz has been working for the Bavarian State Ministry of Environment and Consumer Protection since 2008—mainly in the unit Sustainability in Business and Local Authorities. Her work areas are e.g. the Bavarian Sustainability Strategy and the consultancy of Supervisory Board members where among other things the Bavarian state has the majority shareholding. Between 2002 and 2008, she studied business economics in Germany, Italy and the USA and wrote her master's thesis about sustainability at the Munich Airport GmbH.

Prof. Dr. Klaus Mainzer studied mathematics, physics and philosophy (1968–1972); Ph.D. (1973) and habilitation (1979); assistant and lecturer (1973–1979) at the University of Münster; Heisenberg-scholarship (1980); 1981–1988 professor for philosophy (foundations of exact sciences) at the University of Constance, vice president of the University of Constance (1985–1988); 1988–2008 chair for philosophy of science; dean (1999–2000); director of the Institute of Philosophy (1989–2008); and founding director of the Institute of Interdisciplinary Informatics (1998–2008) at the University of Augsburg; since 2008, he is chair for philosophy of science and technology, director of the Carl von Linde-Academy (2008–2015) and 2012–2014 founding director of the Munich Center for Technology in Society (MCTS) at the Technical University of Munich, member of several academies and interdisciplinary organisations (e.g. The Academy of Europe/Academia Europaea in London, European Academy of Sciences and Arts in Salzburg, National Academy of Science and Engineering (acatech) in Berlin/Munich, Daimler-Benz Foundation, BMW Foundation), president of the Japanese-German Society of Integrative Science (acatech), co-editor and board of scientific journals: e.g. International Journal of Bifurcation and Chaos in Applied Sciences and Engineering (Singapore: World Scientific); Advanced Science Letters (American Scientific Publishers); Kant-Studies (Berlin: De Gruyter); guest-professor in Brazil, China, India, Japan, Korea, Russia, USA, and the European Union.

Anastassia Makarieva holds position of senior researcher in the Theoretical Physics Division of Petersburg Nuclear Physics Institute in Russia, where she works since 1996 when she graduated as biophysicist from the Leningrad Polytechnical Institute in Russia. Anastassia also holds a degree in Scandinavian philology and literature from St. Petersburg State University (1998). In 2000, Anastassia got her Ph.D. in atmospheric physics. Together with Victor Gorshkov in 2007, Anastassia formulated the concept of the biotic pump of atmospheric moisture. In 2008, Anastassia got the L'Oreal-UNESCO-Russian Academy of Sciences Prize "For women in science". Currently, Anastassia's research focuses on quantifying the impact of forests on the continental water cycle. She is also interested in developing a cross-disciplinary theoretical framework for assessing the problem of environmental stability on Earth.

Ortwin Renn serves as full professor for Environmental Sociology and Technology Assessment and as Dean of the Economic and Social Science Department at the University of Stuttgart (Germany). He directs the Stuttgart Research Center for Interdisciplinary Risk and Innovation Studies at the University of Stuttgart (ZIRIUS) and the non-profit company DIALOGIK, a research institute for the investigation of communication and participation processes in environmental policymaking. Renn also serves as Adjunct Professor for "Integrated Risk Analysis" at Stavanger University (Norway) and as Affiliate Professor for "Risk Governance" at Beijing Normal University. Together with Armin Grunwald, he co-directs the German Helmholtz-Alliance: "Future infrastructures for meeting energy demands. Towards sustainability and social compatibility."

Ortwin Renn has a doctoral degree in social psychology from the University of Cologne. His career included teaching and research positions at the Juelich Nuclear Research Center, Clark University (Worcester, USA), the Swiss Institute of Technology (Zuerich) and the Center of Technology Assessment (Stuttgart). Among others, he is a member of the Scientific Advisory Board of EU President Barroso, the Scientific and Technical Council of the International Risk Governance Council (IRGC) in Lausanne, the National Academy of Disaster Reduction and Emergency Management of the People's Republic of China and several national and international Academies of Science. In 2012, he was elected president of the International Society for Risk Analysis (SRA).

His honours include an honorary doctorate from the Swiss Institute of Technology (ETH Zurich), an honorary affiliate professorship at the Technical University Munich, the "Distinguished Achievement Award" of the Society for Risk Analysis (SRA) and several best publication awards. In 2012, the German Federal Government awarded him the National Cross of Merit Order in recognition of his outstanding academic performance. Renn is primarily interested in risk governance, political participation as well as technical and social change towards sustainability. Renn has published more than 30 books and 250 articles, most prominently the monograph "Risk Governance" (Earthscan: London 2008).

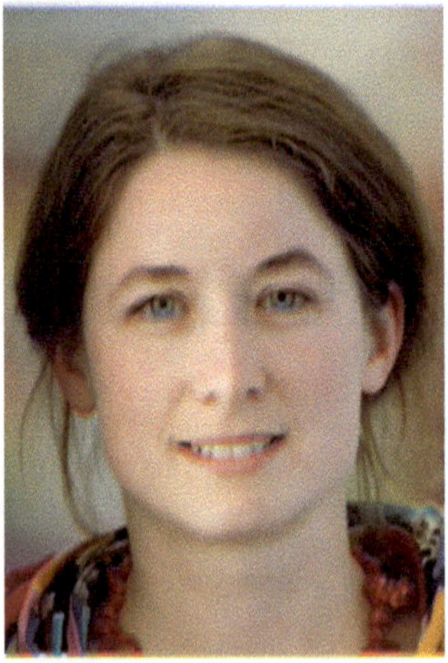

Verena Risse studied law and philosophy at the Universities of Cologne, Paris I Panthéon-Sorbonne, and at the London School of Economics. She is about to complete her Ph.D. in philosophy at Frankfurt University. Since 2013, she also works as a researcher at the Munich School of Philosophy. Her research covers the areas of legal and political philosophy focusing on topics such as global justice, human rights and migration.

Bernhard Schätz received his Ph.D. and Habilitation degree in Informatics from the Technical University of Munich. At the fortiss Transfer Institute associated with the Technical University of Munich, he leads the research department "Software & Systems Engineering" with the fields of Analysis and Design of Dependable Systems, Optimised Design Space Exploration and Model-Based Engineering, with Smart Grid, Automotive and Automation as fields of application. Besides his scientific activities, he acts as lecturer at the Technical University of Munich, is co-founder and member of the advisory board of the Validas AG, and works as a consultant (incl. BMW, Bosch, Eurocopter) in the field of Software and Systems Engineering.

Joachim A. Schütter was educated as a civil engineer at the University of Karlsruhe, Germany. Between 1993 and 2002, he directed the unit Water Construction at the State Office of Watermanagement in Hof, Bavaria. Between 2003 and 2013, he was assistant at the Ministry of Environment, Munich, dealing with international Water management, Water Protection and Regional Planning. In 2009, he graduated in the subject Knowledge Management. Since 2013, he is chair of the unit Water Protection of the Regional Office of Environment, Augsburg.

Katharina Stöckl studied Biology at the Technical University of Munich and at the University of Melbourne, Australia. Since 2012, she works as a researcher at the Chair of Aquatic Systems Biology at the Technical University of Munich. In her doctoral thesis, Katharina is developing conservation management strategies for endangered freshwater mussel species. She also works for the Bavarian coordination office for freshwater mussel conservation, which is financially supported by the Bavarian State Ministry of the Environment and Consumerism and the Bavarian Environmental Protection Agency.

Bernhard Stöckle studied Sustainable Resource Management and Forestry at the Technical University of Munich. From 2004 to 2011, he worked as a scientific employee at the chair of Zoology of the Technical University of Munich and is currently working as post-doc at the Chair of Aquatic Systems Biology of the Technical University of Munich.

Michael von Hauff is a full Professor of Economics and Economic Policy at the Technical University of Kaiserslautern. He studied economics at the Augusta State University/Georgia, USA, and at the Universities of Stuttgart and Constance. He received his Ph.D. (Dr. rer. pol.) in 1977 and completed his postdoctoral habilitation at the University of Stuttgart in 1987. In 1995, he was a visiting professor to the University of Delhi/India. Since 2003, he has been a frequent guest lecturer at the Nanyang Technological University in Singapore and at the Institute of Economics/Yangon Myanmar. Prof. von Hauff is a member of the editorial boards of several international journals, such as "Progress in Industrial Ecology—An International Journal" and "International Quarterly for Asian Studies". Moreover, he is a member of several national and international organisations including the European Academy of Sciences and Arts.

Christian Werthmann is a Professor at the Institute of Landscape Architecture, Leibniz University Hannover. Werthmann accumulated extensive professional and academic experience in Europe and the USA. In his academic work, he researches the implementation of ecological infrastructure in heavily urbanised areas, especially in the non-formal cities of the Global South—a line of research that he initiated as an Associate Professor at the Harvard Graduate School of Design. His most recent research is concerned with the redevelopment of urbanised waterbodies in Sao Paulo and pre-emptive urbanisation strategies for the landslide-prone hills of Medellin in Colombia.

Peter A. Wilderer serves as chairman of the International Expert Group on Earth System Preservation (IESP), a network of scientists, entrepreneurs, administrators and politicians. Peter was educated as an environmental engineer at the University of Karlsruhe, Germany. Between 1982 and 1990, he taught at the Technical University of Hamburg-Harburg. Between 1991 and 2004, he held the chair of Water Quality Control and Waste Management of the Technical University of Munich. In 2003, Peter received the Stockholm Water Prize. He holds two honorary doctoral degrees. The Technical University of Munich honoured him the title of Professor Emeritus of Excellence.

Introduction

Scientific knowledge is the essence of sustainable development, but knowledge remains useless if politically impractical and lacking of social support. To get sustainable development materialise, it appears fundamentally necessary to transfer practicability of scientific knowledge into the mindset of the society. This is a task which has already been discussed centuries ago by the Greek philosopher Plato and his dialogue partners, Glaucon and Socrates. In his book entitled "Republic", Plato explained his concept of thinking by referring to what has become known as "The Analogy of the Sun". The Sun is treated as a metaphor for the nature of reality and knowledge. In analogy to the light emitted by the Sun, intelligent awareness and visibility of knowledge is lit up by truth and reality. Plato writes that it is goodness which provides truth and allows people to gain knowledge (http://en.wikipedia.org/wiki/Analogy_of_the_Sun).

Concerning sustainability wouldn't it be a stroke of luck if goodness expressed by sustainable behaviour were the driving force of sustainable development? Geometrically definable and organised material structures could resemble the nature of reality and knowledge lit up by truth and reality. There are many examples which support this concept. In ancient times, the immunisation against external foes led to the formation of settlements and fortified cities. The transfer of psychological and sociological necessities took place in geometrically defined forms and structures (Wittfogel 1957). The Internet, physically formed of wires and computers, seems to foster globalisation.

This leads to the core question addressed in this book: Does a calculated mixture of centralisation and decentralisation result in an organised and tangible image of sustainability and create a geometrically definable form of being out of which sustainability and resilience can develop almost by themselves?

Decentralised systems are frequently used as synonyms for sustainability: regional economic cycles, food from the region, adapted technologies such as small sewage treatment plants, citizens' identification with their drinking water and local generation of energy. Biodiversity and the diversity of civilisation also seem to match this decentralisation concept. Centralised approaches, on the other hand, promise a high level of efficiency, central warehouses, central management and rules, large sewage treatment plants and energy suppliers. In the following, these theories will be addressed and investigated from different angles.

The contents of the various chapters circle around questions such as:

- Which strategy should be chosen to solve the threatening problems humankind is facing in the twenty-first century?
- Is globalisation a promising way to go, or should we better take the route towards decentralisation?
- Nature works with redundancies and diversity: is this a model for our civilisation?
- Is "Centralised" a synonym for efficiency and "Decentralised" a synonym for resilience? Or does the term "centralised" resemble the properties of community and "decentralised" refer to egoism and greed for profit?
- Does the megacity stand for centralisation and rural development for decentralisation?
- Is "happiness" more correlated with centralised or with decentralised structures?
- Can a basic organisational structure be imposed/promoted by sustainable regional planning (for instance by developing rural areas)?

In the foreground of all those aspects, the port of departure is whether learning from Nature is reasonable question.

Over the past 2.4 billion years, natural systems, ecosystems in particular, were exposed to devastating events (e.g., earth quakes, outbreak of volcanoes, strikes by celestial objects, changes of solar radiation), but life persisted, nevertheless. Presumably, life persisted thanks to auto-regulative mechanisms embedded in living organisms and ecosystems. Collapse was avoided because living systems were able to continuously adapt to changing environmental conditions. We call this ability "resilience".

Other than in anthropogenic systems natural adaptation is not initiated by free will or wisdom but by natural control mechanisms such as wild fire, heavy storms, flooding and droughts. Such mechanisms work because biotic systems are in possession of a variety of options, namely diversity and redundancy of species. Ecosystems are known to be cruel. Species which are not able to adjust get sacrificed. Conservation of the status quo is unknown in nature. Unknown are also control mechanisms resembling social welfare, grants of subsidies to companies prone to bankruptcy and compensation for loss of property after—as humankind calls it—natural catastrophes.

Ecosystems are not uniformly composed on this planet. Species distribution, quantitatively and qualitatively, are adjusted to the very environmental conditions at the spot. Ecosystems are decentralised with respect to their function. For instance, an aquatic ecosystem in mountainous areas differs greatly from such systems in the lowlands. Ecosystems in the soil of Sub-Sahara Africa differ greatly from those in Bavaria or elsewhere. Does this mean that globalisation is unknown in the Nature? *Homo sapiens* is certainly an integral part of Nature and of regional ecosystems as well. However, a number of properties and capacities distinguish humankind from all the other biota. Cognition, rational thinking and transfer of knowledge over generations but also egoism, insatiableness and striving for profit on the expense of others are some of the distinguishing factors. In comparison, the factor "time" plays a very unique role, expressed in the slogan "time is money". Anthropogenic systems are superior with respect to efficiency and control.

Thus, a direct transferability of natural control mechanisms appears hardly possible. Anyhow, we have to admit that the existence and well-being of biota including humankind depend on avoidance of collapse and maintenance of the resilience alike. Based on the discussion above, it appears to be reasonable to consider the function of ecosystems as a model for keeping anthropogenic systems resilient. The contribution of Wolfgang Haber scrutinises this assumption.

Since continuous adaptation to changing external and internal conditions is the essence of resilience, decentralised structures appear particularly appropriate for the solution of problems related to the basic needs of people (supply of water, food, energy, health care, education, equal living and working conditions, maintenance of cultural heritage, etc.). In contrast, centralisation is necessary for proper management and control of technical, economic and governmental structures. In this context, management and control duties include activities keeping systems in the state of continuous adaptation. Again, such assumptions are subject of critical investigation by the authors of the following chapters.

Munich, Germany Peter A. Wilderer
Muenchen, Germany Martin Grambow

References

Plato (390–370 BCE) Republic; Book VI (507b–509c (2007). The Allegory of the Sun. Penguin Classics

Wittfogel KA (1957) Oriental despotism: a comparative study of total power. Yale University Press

Chapter 1
Regulation and Control Processes in Ecosystems

Wolfgang Haber and Jürg Bloesch

1.1 Ecosystem Function

Wolfgang Haber

1.1.1 Key Messages

The question whether and to what extent economic and societal systems should be organized by a central or decentral regime is an important one. Natural ecosystems, however, cannot serve as a model to answer this question. They are a-central in character.

1.1.2 What Is an Ecosystem?

The term 'ecosystem' was conceived as a device to explain and understand the organization of the phenomenon 'life' on the planet earth. At present attention is focused on ecosystem functions and their services to humans. But what is an ecosystem, what is meant by using this term?

In the early 1930s, the British ecologist Tansley searched for a suitable word to denote the physical and biological components of an organism's environment,

W. Haber
Technical University of Munich, WZW faculty, Freising-Weihenstephan, Germany
e-mail: wethaber@aol.com

J. Bloesch (✉); Corresponding author for Section 1.2
Alumnus ETH Zürich, Eawag Dübendorf, Switzerland, Zurich, Switzerland
e-mail: juerg.bloesch@emeriti.eawag.ch

P.A. Wilderer, M. Grambow (eds.), *Global Stability through Decentralization?*,
Strategies for Sustainability, DOI 10.1007/978-3-319-24358-0_1

considered in relation to each other as a unit. From the proposals he received, Tansley chose 'ecosystem' which he defined as follows (emphasis W.H.):

> Though the organisms may claim our primary interest, we cannot separate them from their special environment, with which they form one physical [!] system. . . . It is the systems so formed which, from the point of view of the ecologist, are the basic units of nature on the face of the earth. . . . These *ecosystems*, as we might [!] call them, are of the most various kinds and sizes. . . . They form one category [!] of the multitudinous physical systems of the universe, which range from the universe as a whole down to the atom. (Tansley 1935; Haber 2004/2011)

Tansley called the ecosystem 'a mental isolate', and he never applied it in his research. There were U.S. ecologists who since the 1940s made the term a successful research subject and gave rise to a particular branch of ecology, proving the high heuristic importance of this mental construct (Odum 1973; Golley 1993; Haber 2004/2011). In this way, ecosystem was established as a mainly functional concept and model capable to explain the 'performance' of life in nature. 'Ecosystem function' is, properly speaking, a tautology. The internationally agreed 'Millennium Ecosystem Assessment' (Millennium Ecosystem Assessment MEA 2005) with its concept of 'Ecosystem Services' has raised 'ecosystem' to new importance, and the study of 'The Economics of Ecosystems and Biodiversity' (TEEB 2010) links it with biodiversity as another key issue of today's environmental policy.

1.1.3 Ecosystem Structure and Organization

Following Tansley's definition, the ecosystem model connects living beings (organisms) with their physical environment providing them with life-supporting resources, in particular energy and materials. Their utilization by organisms is based on the property of life as an input-throughput-output system (metabolism) with specific transformations of energy and matter, which are always ending in residues. Material residues are called detritus, comprising also dead or useless parts of organisms and their corpses, but retain their property of a chemical resource which can be re-used and thus recycled. Contrary to this, the energy used in metabolism is degraded to heat which of course still promotes life processes, but cannot be re-used and gets lost; thus life depends on a continuous input of 'fresh' energy (exergy) (Dincer and Rosen 2007).

The organisms utilize these resources according to the principle of division of labour. In order to understand this principle and the systemic organization derived from it, ecologists divide all organisms, notwithstanding their huge abundance and diversity, into only three functional classes called producers, consumers, and decomposers, which at the same time form the three living compartments of an ecosystem (Fig. 1.2).

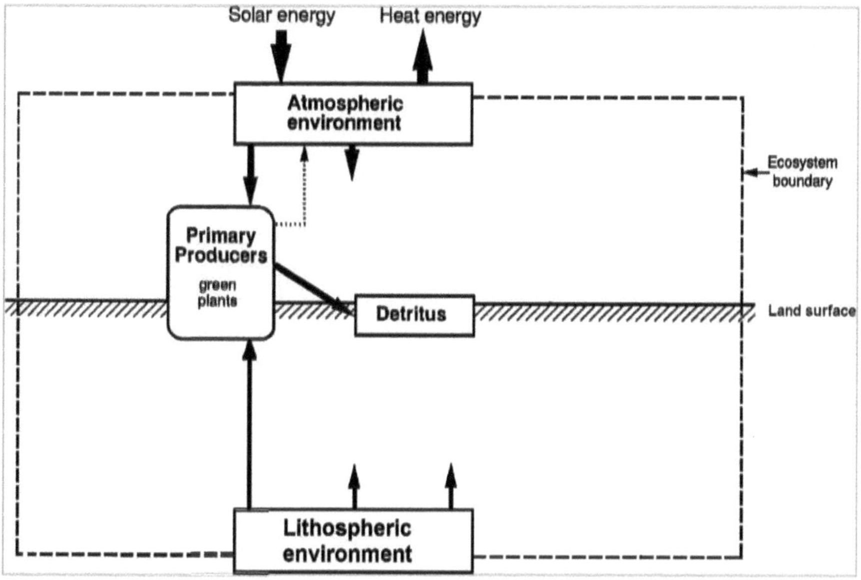

Fig. 1.1 The development of a terrestrial ecosystem begins with the establishment of green plants representing the functional class of primary producers, within the framework provided by the atmosphere and the lithosphere, both permeated with water from the hydrosphere

1.1.3.1 Class of Producers

Producers are embodied by green plants, the only organisms capable to capture solar energy and to fix it by photosynthesis, using carbon dioxide (CO_2) and water, in sugar as the first product of a long sequence of other energy-rich compounds. The plants use them in a double way: firstly as 'life stuff' or 'food' for their own living, secondly for building their own body structure or 'plant architecture' made up (for land plants) by stems or trunks, leaves and roots. The roots which provide plants with water and dissolved minerals break up the substrate and make it accessible for life. Owing to these properties, plants are the essential basis both of life in general (except bacteria living on chemical energy) and of every ecosystem development which begins with plants (Fig. 1.1). As plants are fixed on place and not mobile, they are also used to characterize ecosystems by giving them popular names like forests, prairies, savannahs or peat mires. The peculiarity of plant organisms—referring to this workshop's topic—is their internal organization functioning without any steering or control centre.

1.1.3.2 Class of Consumers

The second functional class of ecosystems, called consumers, is unable to produce their food themselves and obliged to procure it—as 'predators'—from other organisms. This requires particular abilities and a completely different biological organization and structure. Consumers have to actively search for and recognize their food, which requires mobility and a sense of both orientation and discernment. For this, they are equipped not only with special sensory organs, but also with a steering and control centre in the form of a central nervous system culminating in a brain. With regard to this workshop's topic: it is in this organism class where biological evolution has established what we call a centralized organization. Consumers are represented by a large percentage of animals including humans who biologically are mammals (see below Sect. 1.1.5). To ensure, and even improve their food procurement, consumers not only eat plants as producers, but also other consumers, thus establishing food chains or food networks within and between ecosystems. For this mode of life, consumers as a rule are aggressive and have to attack, hurt and even kill the food-providing organisms called 'prey'. Every kind of prey, however, has a survival instinct and tries to escape or to defy the consumers' attacks—which again must be overcome by the food claimers. All this they can only achieve by a central steering capacity.

1.1.3.3 Class of Decomposers

As mentioned before, both plants and animals regularly produce metabolic residues and finally end up as dead bodies. This lifeless organic matter provides the existential basis of the decomposers as the third functional class of ecosystems, represented by small animals and chiefly by fungi and bacteria—mostly living in the soil, hardly visible and therefore often disregarded. For the ecosystem function, they play a crucial double role. Firstly, by using the dead organic matter as food, they split it up into its inorganic components like carbon dioxide or nitrate which are being re-used by the plants in their production of new 'life-stuff'. This fulfils the principle of matter recycling (Fig. 1.2, below left). Secondly, decomposers, chiefly fungi and bacteria, transform parts of the organic matter into a totally new group of chemical compounds called 'humus' (ibid., below right). It is humus which constitutes the fertile soil and the foundation of all biological productivity sustaining life on the earth's continents. Thus, decomposers create their own specific environment. They are much less mobile than consumers, as the detritus they need as food 'comes from above' and has not be searched for or even hunted; thus most decomposers have only a reduced nervous system (as animals) or lack it completely (as fungi and bacteria).

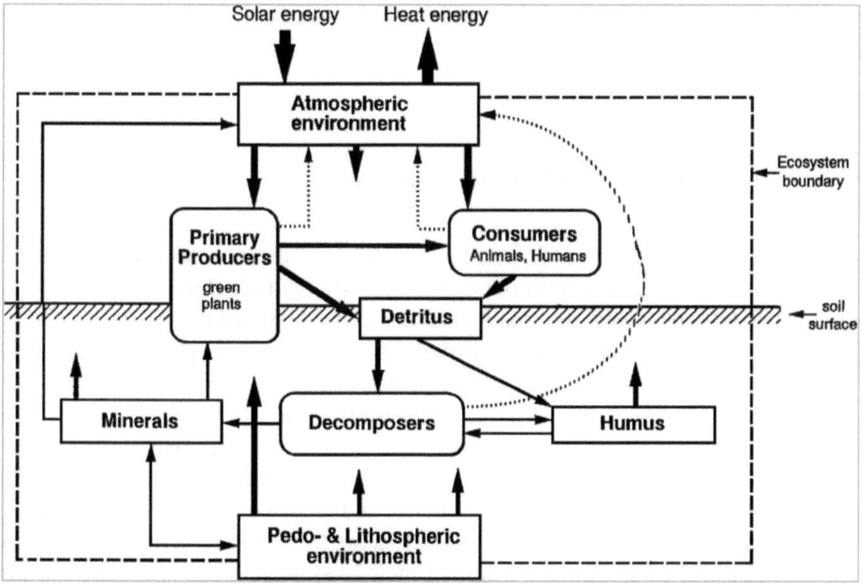

Fig. 1.2 With the functional classes of consumers and decomposers, the ecosystem is complete, functioning with energy flow and matter cycling, but without any controlling centre

1.1.4 Ecosystem Functioning and Regulations: Without Central Control

With these three classes of organisms in localized connection, an ecosystem is functionally complete (Fig. 1.2), mostly with high organizational complexity. The three classes are variably composed of great numbers of different species, with an almost tenfold increase from producers to consumers, and even more with decomposers. A handful of good soil contains more organisms than the global human population. I refrain from describing the manifold regulation procedures needed to keep an ecosystem in permanent function. The most important process is competition among organisms for the resources they need, as these are unevenly distributed or accessible both in time and space, and often scarce or finite. Competition results in winners and losers, effectively regulating population numbers and propagations. Of course there is also cooperation and mutual dependence: with photosynthesis, plants 'exhale' oxygen needed by the animals, and these conversely exhale CO_2 which plants require. Many flowering plants depend for their reproduction on animal pollinators, and often their seeds are spread out by animals.

The most striking fact is that natural ecosystems function, and regulate all processes without any central agency governing and controlling them. This is all the more astounding as the organisms of one of their functional classes, the consumers, do have a control and steering centre for their behaviour—but the ecosystem as a whole can get along without it, in a dynamic and adaptable manner and continuous performance, with only temporary steady states.

1.1.5 The Arrival of Humans in Nature

Within these highly developed, functionally well-organized terrestrial ecosystems, evolution generated, a few million years ago, the hominids (humans) as a unique new organism type (Cook 2005): a mammal endowed with additional assets that no other organism owns: cognitive abilities like intellect, foresight, and conscious feelings, capable to assign values to all objects they perceive. This evolutionary event was only possible within the ecosystem class of consumers which are equipped with a steering organ, in the case of primate mammals with a brain, which for humans became their intellectual centre. With its capabilities and growing capacities, humans were able to recognize and to assess the natural environment surrounding them, and to learn how to get a secure footing in it, in order to not only survive, but also to continually improve their livelihood. Of course they did not know anything about ecology and ecosystems, but they must have developed an intuitive notion and motivation for using their intellect to master and control nature—instead of only adapting to it.

My thesis is that early humans, in the course of thousands of generations, have intuitively become aware of the organization of their natural environment and developed the idea of initiating their own, cultural evolution. They wanted to establish a specific human-made environment, shaped and arranged according to purely human interests and values, implanted in (or 'grafted' on) the natural environment, aiming at its domination. In retrospective, and related to today's ecological knowledge, humans have largely succeeded with this evolutionary goal.

As main driving forces of humans, which all other living beings lack, I identify

- The constant search for technical solutions of problems, aided by new energy sources;
- The steady quest for 'more'—better, faster, easier, healthier, higher, longer, happier ... in quantity as well as quality; and
- The conscious preservation of each human individual as humanitarian duty.

Humans are the only organisms being aware of individual death and collective extinction, which they by all means wish to delay or avoid, respectively. These human characteristics, however, are not compatible with the functional principles of natural ecosystem organization, and they even raise conflicts within humans themselves, who as biological beings have evolved in the ecosystem class of 'aggressive consumers' being predisposed to injure or to kill. This is happening until today, often cerebrally enacted, and not excluding other humans.

1.1.6 Formation of the Human Environment

The first main step of creating a human environment was supplementing solar energy with using fire as a by far much stronger energy source, disposable as desired everywhere at any time for many purposes, enhancing human power, food range, and influence into far-reaching dimensions. Fire, however, requires fuel whose first principal source humans discovered in wood, creating a new and lasting dependence on trees and shrubs, and it is extremely dangerous, thus created new risks and hazards. But without fire, humans would have been unable to spread from their tropical origin across all continents into colder climates with long dark winters, and to develop many technologies like metal use.

The second main step of human evolution, only 10,000 years ago, was changing human food supply from gathering and hunting to agriculture, in particular crop farming with cereals. This fundamental transformation forced humans, turned into farmers, to radically eliminate the natural plant cover with the aid of stone tools and fire, and to dig up (and seriously impact) the soil for sowing or planting their crop plants. After a few centuries, crop farming became the principal human food source, supplying staple food, supplemented by livestock farming which, however, can never produce the bulk supply. With this bio-technical transformation, humans created their first real own environment at the expense, and to the detriment of the natural environment whose components were strictly kept apart from it, even fought and locally exterminated. Moreover, humans now made a clear distinction between a 'wild' and a 'domesticated', cultivated nature. As human numbers increased, owing to more and better food, cropland expanded and continually displaced original natural ecosystems.

But this was only one side of the transformation. Farming obliges humans to a more sedentary way of living, thus they had to construct solid, long-lasting houses supplemented by granaries, livestock stables, workshops, water-wells, waste pits, and transport roads. Before long, farmsteads grew into villages needing more building grounds. Compared to crop farming, building upon the ground destroys natural land cover completely including the soil, which when lost cannot be restored.

1.1.7 Rural and Urban Living: The Anthroposphere

Since the invention of agriculture, humans live in and from two artificial systems which do not occur in nature (Manning 2004). One produces their bulk food and consists of natural organisms, mostly of a single species or variety in uniform, short-lived cultivation, but is completely depending on human control and management. Left to itself, a crop field would soon disappear. The other system, created to accommodate humans with their equipment and infrastructure, consists of dead, solid materials and structures. The materials are of natural origin, but the structures are alien elements in living nature and destroy it almost completely.

In the course of centuries, both systems, called 'rural' and 'urban', depending on each other, have been expanded and intensified, supported by inventions like vehicles with wheels and animal traction. The metal age, made possible by using fire, brought new, more effective tools and techniques. The farmsteads grew into villages and townships, the farmers' fields, pastures and forests had to be adapted to the growing urban supply. Bogs, swamps and floodplains were drained, wilderness eliminated. With the introduction of science-based methods for farming and land use, and entering the industrial age, artificial fertilizers replaced livestock dung, machines driven by fossil fuels or electricity substituted human and animal physical work, human livelihood became a chemo-technical enterprise.

Urban living and culture, depending on farming, is attracting a steadily increasing number of people, thus reducing but at the same time challenging the rural population, as fewer and fewer farmers have to produce more and more rural products for satisfying growing urban needs. At the same time, the spatial expansion of cities and traffic is reducing croplands and pastures. Moreover, both urban and rural human systems are negatively affected by the harmful and destructive side- and after-effects of all these techno-industrial innovations and activities which, however, are developed to improve human well-being (Jay 2000). On the global scale, all these problems are aggravated by the growth of the human population both in numbers and in demands, and by their political and cultural differences (Wackernagel and Beyers 2010).

With this cultural evolution, which, by the way, is characterized by much continental and even regional diversity, humans have succeeded in establishing their own 'artificial' environment, often referred to as anthroposphere or noosphere, within the terrestrial biosphere. But they achieved this without any scientifically founded knowledge or understanding of the biosphere's or the earth system's natural organization and function. Moreover they ignored, or did not realize that, and how the human-made environment, with all its technical perfection and centralized management, is not only connected with the earth system's 'wild' nature, but functionally depends upon it, left apart its aesthetic attractions. Only in the last 100 years, with the rise of the young science of ecology (Küster 2005; Haber 2007b), humans have become aware of that crucial knowledge deficiency and of their responsibility for the preservation of the earth system. In comparison to several millennia of humans' technical and cultural progress, 100 years are too short a time for establishing, and agreeing upon a reliable new road into a more sustainable future with fundamental changes in the human-nature-interrelationships.

1.1.8 Natural Ecosystems Compared to Human-Made Systems

Coming back to the ecosystem concept whose functions and services, according to the Millennium Ecosystem Assessment, are to benefit and improve human well-being (MEA 2005). It is still debated if the term 'ecosystem' is preferably or exclusively related to natural ecosystems, as indicated by Daily (1997), or includes cultivated ecosystems like crop fields, pastures or managed forests which fulfil indispensable supply or provisioning services for human livelihood. These cultivated systems, however, only partly correspond to the ecosystem model with its three functional classes of producers, consumers and decomposers as described above (Fig. 1.2). In a natural ecosystem the three classes occupy, and operate on the same site or place in self-regulation. The 'system wheat field', on the other hand, consists only of producers, reduced to one species or variety (other producer species being eliminated as competitors). Its human consumers are elsewhere in urban settlements, and the on-site animal consumers which feed on wheat plants are eliminated too. The decomposers in a wheat field are tolerated by the farmer, though disturbed by soil treatments, but the harvest reduces their regular supply of dead organic matter they need as food, and use for humus formation. Compared to a natural ecosystem, a wheat field is only a 'torso ecosystem' and cannot maintain itself.

By the way, the human invention of agriculture has in a certain manner, but intuitively imitated the natural ecosystem principle of division of labour (Fig. 1.3). Green plant producers were replaced by crop fields, consumers were split up into a food chain from livestock to humans, the livestock kept on pastures, in large stables or on large feeding units, humans in likewise large artificial urban centers. The decomposers, for long-time unknown or neglected, were discovered when organic waste and residues from urban peoples' metabolism accumulated to unbearable masses and required special treatment. For this purpose, townspeople installed 'decomposer cultures' in sewage treatment or compost plants. In this sequence, the trias of producers, consumers and decomposers was kept in function, but the three classes were dislocated to separate, sometimes rather distant places, requiring new transport systems because they have to remain connected.

1.1.9 Conclusion

Referring to the topic of this book, the conclusion I draw from these ecological considerations is that a natural ecosystem cannot serve as a model for human-made systems, be they centrally or decentrally organized. As I described above, natural ecosystems function, and develop themselves, without a steering or control centre or an agency. They are truly 'a-central'. In contrast to them, humans' cultural environmental systems, implanted in the natural systems, cannot function without

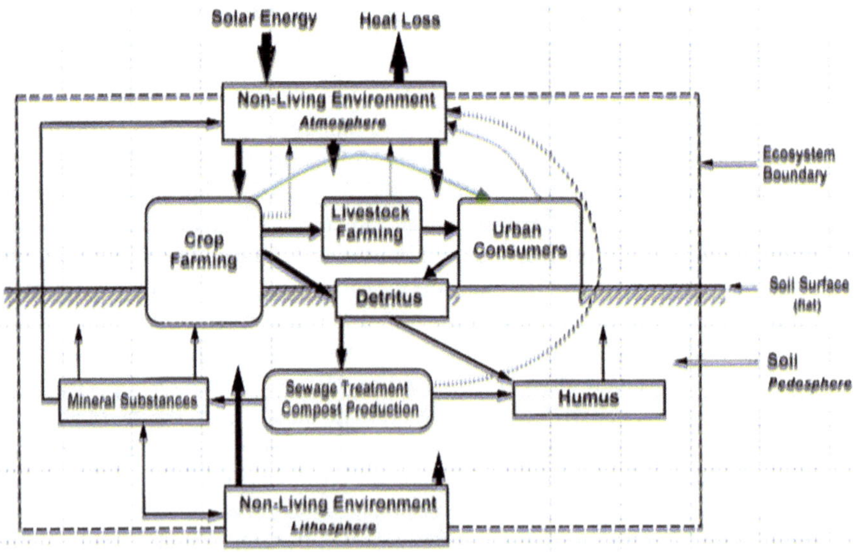

Fig. 1.3 Artificial system compartments created and controlled by humans have replaced the natural functional classes (cf. Fig. 1.2), causing continuous ecological conflicts

a central control agency. Humans, evolved from the ecosystem class of consumers equipped with a brain as steering organ, are intellectually and socially unable to fit themselves in an a-central earth system, much less to submit to it. Moreover, following humanitarian obligations, humans exempt themselves from many natural ecosystem regulations, such as limiting population sizes. Even in the early long period of gatherers and hunters, humans organized themselves in small social groups with a central head or leadership. And when we today create nature reserves where nature is left to itself in a-central ecosystems, we apply a centrally organized control and management, as exemplified by the Habitats Directive (FFH) of the European Union (Haber 2007a).

Humans have to continue living with centrality, though in quite different organization structures ranging from large, globally influential centers to small local centers, within an earth system which has no central agency—but humans seem to tend to establish it, intuitively or intentionally. The MEA proclamation of ecosystem services to promote human well-being derived from well-being of 'wild' nature, but superior to it, may be seen as an example (Haber 2014). Its implementation is doubtful.

1.2 Globalization and Balance: The Ecosystem Dimension

Jürg Bloesch

1.2.1 Key Messages

To master the future global environmental problems balancing things must become the core of the political agenda. Ecosystems teach us that a balanced state or better to say a dynamic equilibrium (homeostasis) is characterized by continuous change. Extremes are not stable and politically dangerous. The “best” balance between extremes such as centralized and decentralized solutions or systems has to be based on science, philosophy, ethics, social welfare, and even on new economic theories. The clue to solve our environmental crisis definitively lies in a fundamental change of the globalized economy. If the long-term trend of growth cannot be stopped humankind rather than the global ecosystem will perish since humans are subject to ecosystem function despite their technology.

1.2.2 Introduction

It has to be emphasized that decentralization, centralization and globalization are anthropocentric views of system organization. Both decentralized and centralized human systems have an organizational center. While decentralized systems are small and local, centralized systems cover large areas up to the global scale requiring technical transport systems. Scaling is important: Decentralized systems function closer to autarky and self-organization than centralized systems.

In search for the right balance between centralized and decentralized solutions, we need first to understand our basis of life (i.e. balanced natural ecosystems, nature; Schwarz and Jax 2011), then to analyze its major impacts (i.e. unbalance by human technology and the power of economy), and thirdly to find optimum measures between dangerous extremes. A general paradigm change is necessary: human growth and overexploitation of natural resources must be stopped, business as usual and economic excesses must be abandoned.

1.2.3 The Role and Function of Natural Ecosystems

According to the evolution of planet Earth and its biosphere, natural ecosystems are self-organized, self-regulated, resilient, and without a steering center (see Sect. 1.1,

Haber). This applies also to the sessile components (basically plants), while motile organisms (basically animals) have developed a steering center in form of a central nervous system and the brain of vertebrates and ultimately Homo sapiens.

Organisms form populations and complex ecosystems, e.g. food webs and predator–prey relationships. There are numerous ecological niches that can be used by various specialized species that need small or large areas for living. There are also ubiquitous species tolerating wider ranges of chemical-physical boundary conditions. Small and large ecosystems are intertwined in a large-scale network. This prevents inbreeding effects in isolated populations by enabling the exchange of genes. Moreover, zoogeographic distribution patterns include long-distance migration (e.g. insects, fish, birds, mammals) aiming to optimize the survival of individual species. A great variety of survival strategies is generally observed with plant and animal species. All species have their distinct role in maintaining basic biological functions that drive the flux of matter and energy—primary production of organic substance by plant photosynthesis, consumption of food at various trophic levels, and decomposition to basic chemical compounds to produce humus/soil on land. Biogeochemical and biological cycles are the basic processes of recycling in ecosystems (Odum and Barrett 2004).

The global climate influences terrestrial and aquatic ecosystems, i.e. the vegetation and water bodies which in turn provide the patchiness of habitats and biota. Landscape features and spatial structural heterogeneity, evolving from, e.g., seasonal and annual cycles (the temporal dimension), stochastic weather events and/or temperature or chemical gradients provide a dynamic system that is the basis for Earth's biodiversity. The resilience and stability of ecosystems are important when considering natural disasters such as earthquakes, volcano eruptions, floods, droughts, fires and epidemic diseases. On a larger scale, there is a balance between pioneer populations, emerging after disasters, and mature climax populations. Similarly, during the global evolution of over some 4.5 billion years, biodiversity was diminished repeatedly by 70–90 % during five significant global events (Raup 1986), but recovered gradually to the new environment. Survival of the fittest, not the strongest, was and is the fundament of evolution, i.e., natural selection (Darwin 1869). Today, we are in the Anthropocene era, and Homo sapiens is the only species dominating the entire planet.

1.2.4 Ecosystem Service and Human Use: Artificial Human Ecosystems

Ecosystem function at all scales is the natural basis for human well-being, activities and development (i.e. human living environment, economy, technology and management, communication and mobility). As such, nature provides ecosystem services that must be clearly separated from the human use of ecosystems.

The Millennium Ecosystem Assessment (MEA 2005) has grouped ecosystem services into four categories: "supporting, provisioning, regulating and cultural services". This anthropocentric MEA definition is being debated as it does not clearly distinguish between services and uses. This border, although not distinct, should be drawn on the application of human technology. For example, an ecosystem service is if a person drinks water from a river or a spring, whereas the installation of drinking water supply systems for communities is a human use. In general, human use by advanced technology rapidly extends into the exploitation of limited natural resources, including land use by agriculture, cities and industries. These exploitations can be called "the ecological traps of humankind" (Haber 2007). Ecosystem services have an intrinsic value (e.g. biodiversity, genetic pools, pollination, oxygen production, carbon dioxide fixation, etc.) and an extrinsic value (such as food provisioning, water purification, groundwater supply, climate and atmospheric regulation, flood/drought mitigation). In contrast, human use is linked with money and price, economy and technology. The political challenge of today is how to propagate intrinsic, non-monetary values of ecosystems and their components (i.e., landscapes, water bodies, animals, plants).

§1 of the Wildbad-Kreuth Declaration (see Chap. 7) states that natural ecosystems cannot be considered as a normative model for human-made systems, as humans add new dimensions and anthropocentric aims by thinking and, hence, tend to dominate natural ecosystems by inventing and using technology. The hypothesis that the human brain (cerebrum) is a failure of evolution (Löbsack 1974) may well be true. Although natural ecosystems provide the basis of human well-being and survival, humans have created and still create artificial ecosystems (e.g. urban areas, reservoirs, agricultural monocultures), globalized frameworks and new technical infrastructures. Many concepts copy biological functions; nevertheless, this undermines the function of vulnerable natural ecosystems. Humans have introduced (1) the spatial separation of three basic ecosystem functions by producers (industry), consumers (in agglomerations) and decomposers (waste disposal) which resulted subsequently in inducing links between them by technological transport systems; (2) socio-technical revolutions by agriculture (replacing the human ecological period of gatherers and hunters), fire and technology; and (3) trade and money fostering globalization and power gains to exploit nature. Overexploitation and inadequate reactions to environmental changes can lead to the collapse of whole societies such as the Polynesians on Easter Island and the Mayas in Central America (Diamond 2005). Although governments have created protected areas (national and regional nature parks, UNESCO-Biosphere reserves, NATURA 2000 areas, Ramsar wetlands, etc), implementation of protection is insufficient as economic pressures are politically stronger despite the worldwide "sustainability" discussion. More political willingness is needed to balance the conflicts of interest and to strengthen the protection of ecosystems, as exemplified in the Arctic. Increasing public awareness with the crucial support of Non-Governmental Organizations (NGOs) may be the first step to improve this critical situation.

1.2.5 Natural and Human-Made-Growth: The Impact of Economy

Birth rates are naturally balanced by death rates ("eat and be eaten") and population growth is limited by biotic and abiotic factors. In contrast, quantitative human population growth is increased by progress in mainstream medicine that aims, as humanitarian obligation, to artificially keep individuals alive as long as possible. Hence, the carrying capacity of natural ecosystems is exceeded. In addition, qualitative growth by higher demands request increasing amounts of industrial food, energy and resources. The human population is highly unbalanced and its exponential growth and demands lead to a dangerous extreme (Fig. 1.4).

Ultimately, humans deprive themselves of their living basis by exploiting limited resources, abusing ecosystems by dumping waste, and inducing global climate change. An unlimited and globalized free market economy is the key driver of excessive growth and regional as well as global environmental problems. The present theories and practice of economy focusing on profit maximization and constant growth are a misconception in view of ecosystem function. Hence, the present primacy of economy as key driver of unsustainable development and exploitation of resources must be broken. The role of humans as global regulators via economic and political arrogance, and by religious fervor, must be radically changed. Human behavior should be focused on saving the GAIA ecosystem "Earth" (Lovelock and Margulis 1974; Lovelock 1995, 2009) sensu Meadows ("The limits to Growth"; Meadows et al. 1972, 1992, 2004) and Capra ("The turning point"; Capra 1982).

Current capitalism, globalization and neo-liberalism will inevitably lead to a dead end, i.e. the breakdown of economic and social systems, since there is simply no eternal growth ("trees do not rise up to the sky"). To break this disastrous trend a simultaneous paradigm change in the domains of economy, politics, social justice and environmental protection is necessary. This is a slow and arduous process that needs specific, responsible and outstanding leadership with a strong ethical background. This is difficult if not impossible, since greed and striving for power are basic attributes throughout human history (Gigantès 2012). At the end, it is about balancing personal and community interests, economic regulation and deregulation, as well as decreasing the discrepancy between the rich and the poor.

The political discussion about sustainable use or development does not help solve the global crisis. There are over 200 definitions (Jucker 2002) and politicians as well as managers use what they like best. While the original definition is based on forest ecology, namely don't use more of the resources than what can naturally re-grow, the most prominently used definitions are the Brundtland definition of the Rio 1992 declaration (leave an intact environment to the next generation, i.e. the grandchildren) and the balance of ecological, economic and social domains. Ecosystem resilience and self-regulation lies behind the concept of sustainability (Bloesch et al. 2015). However, the implementation of sustainability is more than doubtful, as exemplified by unreachable national goals set for mitigating

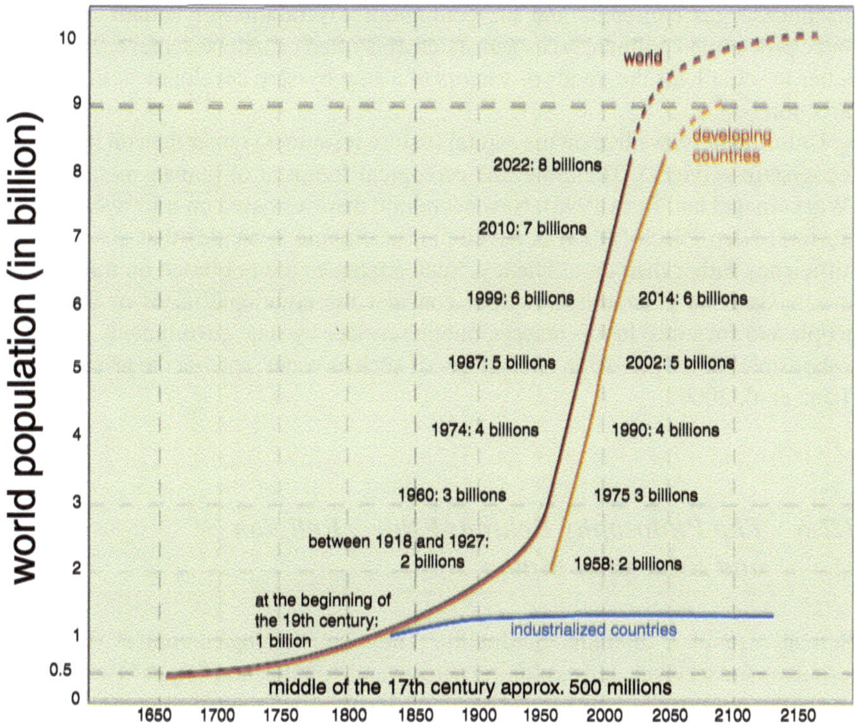

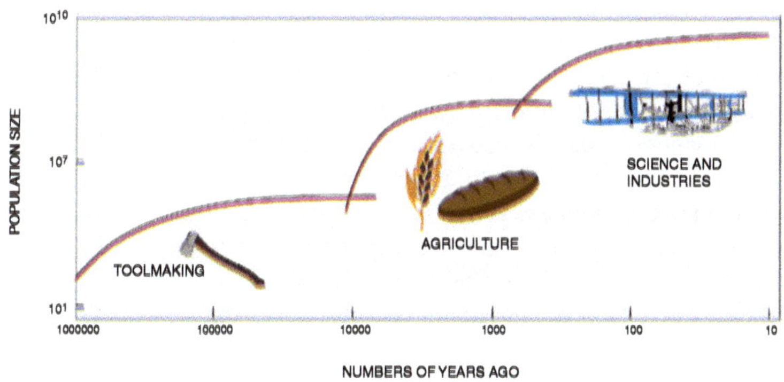

Fig. 1.4 Human growth: when does it stop? While the individual life cycle is closed by birth and death, the large-scale population cycle on a "geological" time scale is not yet closed. (*Top picture*) Exponential human growth and extrapolation into the future. Human decoupling of birth and death rates led to growth increase. The predicted lag phase and subsequent decline are due to decreasing birth rates possibly caused by nature induced loss of fertility. After Leisinger (1994). (*Bottom picture*) Three phases of the long-term development of humankind: (1) the invention of stone-tools allowed a population increase from 150,000 to 5 million people; (2) the agricultural revolution allowed support of 500 million people; (3) the technical and scientific revolution may allow for up to 10 billion people. With every step of development, the direct dependence of natural systems decreased, and periods of strong growth are followed by periods of stagnation that seem to represent the basis for the next growth phase. After Kates (1994)

greenhouse gas emissions and the continuous exploitation of natural resources. Since proclaimed political goals and statements often cannot be achieved, it may be better to set at least the trends or vectors of a step-by-step development in the right direction.

Unfortunately, we live on the capital (nature resources) rather than on its interest (ecosystem services). Therefore, the ecological footprint of humans is far too big (Wackernagel and Rees 1996). New economic theories based on true sustainability are rare. Nico Paech (2012) is developing economic strategies that are based on sufficiency rather than on efficiency. Such strategies are not based on materialism and maximizing economic wins, but consider the emotional needs of individual people and societies. In this respect, public services by state governments should be responsible for the provision of basic goods such as water, and not the private sector (Lanz et al. 2006).

1.2.6 The Philosophy Behind: Ethics, Religion and Education Matter

Humans are in a dilemma of duality. They are "biological units" (living on resources for survival, and exploiting nature) and "intellectual beings" (with ethical and religious concerns to protect nature) (Haber 2013). Natural ecosystems are recognized as basic units of life, scaled up from local/regional ecosystems (e.g. ponds, lakes, streams, landscapes) over large ecosystems such as forests and oceans to the global GAIA ecosystem. In contrast to the widely existing Cartesian view (man conquering nature), and in agreement with, e.g., the conception of Deep Ecology by Arne Naess (1989, 1995), humans are considered as part of nature and hence a component of natural ecosystems, as are other biotic and abiotic components. As explained above, due to a highly developed brain and respective technology, the human population grows virtually unlimited and man has the capacity to create artificial ecosystems and the capability to destroy natural ecosystems. However, man cannot trick concrete and basic scientific theories or fundamental natural laws. For example, gravitational forces (Newton's physics) apply to all, energy can neither be generated nor destroyed but only be transformed (first thermodynamic law), and Einstein's relativity theory affects daily life. The world's mechanism is the principle of waves, i.e. oscillation (the amplitude) between two extremes, reflecting duality or bipolarity ("every coin has two sides"). Therefore, we should work with, not against nature.

Paradigm shifts only happen if they are based in society, i.e. from the heart and by the conviction of many individuals (bottom-up). The unity of body, spirit and soul represents the trilogy of economy, social welfare and economy. Crucial is the individual perception of the sense of life, standard of living and education. As such, a balance between soul, spirit and body should be achieved on an individual basis, while balancing between the basic needs and wealth of societies. Ultimately, this balance also addresses equal rights and democracy. In general, the mostly semipermeable boundaries as a biological principle from cell walls up to ecosystem borders provide the basics for balance.

Education into ecology, biology, social behavior and ethics is of crucial importance and should simultaneously address teachers, children and adults. Therefore, governments should invest in transdisciplinary education systems to use the great potential of human thinking. To better conserve the environment, systems management should apply several principles, e.g. polluter/user pay, best available technique/practice, non-deterioration of ecosystems, precaution, subsidiary, solidarity, transparency (public participation), and an integrated and holistic approach.

1.2.7 Balance Between Centralization and Decentralization in Technical Systems

Human technology should adapt to nature, i.e. generally be decentralized and not centralized. Only due to technology, centralized systems could be created (e.g. waste water treatment plants, drinking water supply, urban drainage systems, economy and global market system, information and communication technology, transport). On a limited regional scale, such centralization has its benefits in terms of efficiency and living standards. Therefore, in reality, we should achieve an optimum balance between centralization and decentralization. Actually, decentralization is coming more in use, e.g. organic farming without using chemical fertilizers and pesticides vs. industrialized farming of (genetically modified) monocultures, and the turning point in energy policy—minergie houses, solar energy vs. charcoal, nuclear, hydropower and wind energy production with an extended storage and transport system.

Usually, technical solutions cannot solve problems truly sustainably but provide only short-term solutions. New technology often needs more energy and resources than it pretends to save. Technical restoration of deteriorated ecosystems depends on scaling: we have restored small lakes, but can we restore the oceans? And can we handle responsibly geo-engineering in space? After some experience, technical solutions often turn out to be a disadvantage causing new environmental problems. For example, waste water treatment plants were designed to remove sewage and nutrients to mitigate pollution, but heavy metals are concentrated in the sludge, and they could not reduce the load of dangerous and persistent organic substances. Only very recently, additional technologies developed for drinking water treatment such as activated charcoal filters and ozonization are being applied. However, technical progress cannot be stopped due to the elemental curiosity inherent in humans. There is hope that cyber physical systems as described in §5 of the Wildbad-Kreuth Declaration (see Chap. 7) can bring a breakthrough so as to stop end-of-pipe solutions and foster strategies that tackle the causes and not the effects.

References

References of Section 1.1

Cook M (2005) A brief history of the human race. W. W. Norton, New York/London

Daily GC (1997) Nature's services: societal dependence on natural ecosystems. Island Press, Washington, DC

Dincer I, Rosen MA (2007) Exergy. Energy, environment and sustainable development. Elsevier, Oxford

Golley FB (1993) A history of the ecosystem concept in ecology. Yale University Press, New Haven

Haber W (2004) The ecosystem. Power of a metaphysical construct. In: Zehlius-Eckert W, Gnädinger J, Tobias K (eds) Landschaftsökologie in Forschung, Planung und Anwendung, Schriftenreihe Landschaftsökologie Weihenstephan, vol 13, Freising, Germany, pp 25–48. [Revised version (2011) An ecosystem view into the 21st century. In: Schwarz A, Jax K (eds) Ecology revisited. Reflecting on concepts, advancing science. Springer, Dordrecht/Heidelberg, pp 215–227 (Chapter 18)]

Haber W (2007a) Zur Problematik europäischer Naturschutz-Richtlinien. Jahrbuch des Vereins zum Schutz der Bergwelt, vol 72. München, Germany, pp 95–110

Haber W (2007b) Energy, food, and land – the ecological traps of humankind. Environ Sci Pollut Res 14(6):359–365

Haber W (2014) Ökosystemleistungen – strapaziertes Modewort oder neues Naturverständnis? Nat Landsch 89(2):50–55

Jay P (2000) Das Streben nach Wohlstand. Die Wirtschaftsgeschichte des Menschen. Ullstein, Propyläen, Berlin. [Edition (2006) Patmos/Albatros, Düsseldorf, Germany]

Küster H (2005) Das ist Ökologie. Die biologischen Grundlagen unserer Existenz. C. H. Beck, München

Manning R (2004) Against the grain. How agriculture has hijacked civilization. North Point, New York

MEA, Millennium Ecosystem Assessment (2005) Ecosystems and human well-being: synthesis. Island Press, Washington, DC

Odum EP (1973) Fundamentals of ecology, 3rd edn. Saunders, Philadelphia

Tansley AG (1935) The use and abuse of vegetational concepts and terms. Ecology 16:284–307

TEEB (2010) The economics of ecosystems and biodiversity. Mainstreaming the economics of nature. A synthesis of the approach, conclusions and recommendations of TEEB. www.teebweb.org

Wackernagel M, Beyers B (2010) Der Ecological Footprint. Die Welt neu vermessen. Europ Verlagsanstalt, Hamburg

References of Section 1.2

Bloesch J, von Hauff M, Mainzer K, Venkata Mohan S, Renn O, Risse V, Song Y, Takeuchi K, Wilderer PA (2015) Sustainable development integrated in the concept of resilience. Problemy Ekorozwoju/Problems of Sustainable Development 10(1):7–14. [see also UN e-Inventory: Sustainable systems in the post-2015 development agenda: Keeping societies, economies and ecosystems resilient. http://www.sdgseinventory.org/searchauthordetails.php?proposalId=261&sdgAuthorId=681. Assessed 24 Nov 2014]

Capra F (1982) The turning point: science, society, and the rising culture. Simon and Schuster, Bantam, Paperback

Darwin C (1869) On the origin of species by means of natural selection, or the preservation of favored races in the struggle for life, 5th edn. John Murray, London

Diamond JM (2005) Collapse. How societies choose to fail or succeed. Viking, New York

Gigantès P (2012) Macht und Gier in der Weltgeschichte. Anaconda Verlag, Köln

Haber W (2007) Energy, food, and land – the ecological traps of humankind. Environ Sci Pollut Res 14(6):359–365

Haber W (2013) Ökologie: eine Wissenschaft unbequemer Wahrheiten – auch für die Ethik. Aus: Vogt M, Ostheimer J, Uekötter F (Hrsg) Wo steht die Umweltethik? Argumentationsmuster im Wandel, S. 325–343. Metropolis, Marburg (Beiträge zur sozialwissenschaftlichen Nachhaltigkeitsforschung Bd. 5, 457S.)

Jucker R (2002) Our common illiteracy – Education as if the Earth and people mattered, vol 10, Serie Umweltbildung, Umweltkommunikation und Nachhaltigkeit. Peter Lang Verlag, Frankfurt a.M., Germany, 346 pp

Kates RW (1994) Sustaining life on the earth. Hope for an environmentally sustainable future lies in evolving institutions, technology and global concern. Sci Am 1994:92–99, Special Issue October

Lanz K, Müller L, Rentsch C, Schwarzenbach R (2006) Wem gehört das Wasser? Lars Müller Publishers, Baden, Schweiz, 536 pp

Leisinger KM (1994) Bevölkerungsdruck in Entwicklungsländern und Umweltverschleiss in Industrieländern als Haupthindernisse für eine zukunftsfähige globale Entwicklung. GAIA 3:131–143

Löbsack T (1974) Versuch und Irrtum. Der Mensch: Fehlschlag der Natur. C. Bertelsmann, Der Mensch

Lovelock J (1995) The ages of Gaia: a biography of our living earth. W. W. Norton, New York

Lovelock J (2009) The vanishing face of Gaia. Basic Books, New York

Lovelock JE, Margulis L (1974) Atmospheric homeostasis by and for the biosphere: the Gaia hypothesis, vol 26(1–2), Tellus, Series A. International Meteorological Institute, Stockholm, pp 2–10

Meadows DH, Meadows G, Randers J, Behrens WW III (1972) The limits to growth. Universe Books, New York

Meadows D, Meadows D, Randers J (1992) Beyond the limits. Chelsea Green, White River Junction, VT

Meadows D, Randers J, Meadows DH (2004) Limits to growth: the 30-year update. Chelsea Green, White River Junction, VT, 368 pp

MEA, Millennium Ecosystem Assessment (2005) Ecosystems and human well-being. Synthesis reports. http://www.millenniumassessment.org/en/Synthesis.aspx. Assessed 24 Nov 2014

Naess A (1989) Ecology. Community and lifestyle: outline of an ecosophy. Trans. David Rothenberg. Cambridge University Press, Cambridge

Naess A (1995) The shallow and the deep, long-range ecology movements: a summary. In: George S (ed) Deep ecology for the 21st century. Random House, Boston

Odum EP, Barrett GW (2004) Fundamentals of ecology, 5th edn. Cengage Learning, Stamford, 624 pp

Paech N (2012) Befreiung vom Überfluss. Auf dem Weg in die Postwachstumsökonomie. Oekom Verlag, München

Raup DM (1986) Biological extinction in earth history. Science 231(4745):1528–1533

Schwarz A, Jax K (eds) (2011) Ecology revisited. Reflecting on concepts, advancing science. Springer, Dordrecht, 444 pp

Wackernagel M, Rees W (1996) Our ecological footprint: reducing human impact on the earth. New Society, Gabriola Island, BC

Chapter 2
Control and Management of Man-Made Systems

Patrick Dewilde, Ulrich Drost, Martin Grambow, Bernhard Schätz, and Martin Korndörfer

2.1 Centralized Versus Decentralized Systems: A Critical Appraisal

Patrick Dewilde

2.1.1 Key Messages

This chapter is devoted to gauge the effects of centralized vs. decentralized organization of dynamical systems. From the analysis it appears that "alignment of intelligence" between agents is key to achieve optimality as well as stability in distributed systems, and this alignment of intelligence is made possible thanks to communication. It appears that in any well-functioning and intelligent system, there shall be an equilibrium between local control and central control, and there shall be

P. Dewilde (✉); Corresponding author for Section 2.1
Technical University of Munich, Munich, Germany
e-mail: p.dewilde@me.com

U. Drost (✉); • M. Grambow (✉) Corresponding authors for Section 2.2
Bavarian State Ministry of the Environment and Consumer Protection, Munich, Germany
e-mail: ulrich.k.drost@t-online.de; martin.grambow@stmuv.bayern.de

B. Schätz (✉); Corresponding author for Section 2.3
fortiss GmbH/department of informatics, Technical University of Munich, Munich, Germany
e-mail: schaetz@fortiss.org

M. Korndörfer (✉); Corresponding author for Section 2.4
Alumnus Technical University of Munich, University of Canterbury, Christchurch, New Zealand
e-mail: m.korndoerfer08@aol.com

P.A. Wilderer, M. Grambow (eds.), *Global Stability through Decentralization?*, Strategies for Sustainability, DOI 10.1007/978-3-319-24358-0_2 rules of interaction between the intelligent agents participating. This brings up the question of ethics, which is defined as the actual behavioral principles of the participating intelligent agents, and the discussion focusses on how desired behavior (desired ethics) can be made concordant with actual ethics. System-wide conditioning plays an important role in achieving this.

2.1.2 Example of Systems

This paper is devoted to systems and in particular dynamical systems, their properties and how they can be designed, with focus on how the choice of centralized vs. decentralized affects those properties. But before engaging in definitions and properties, let us consider some examples of centralized and decentralized systems.

2.1.2.1 Example 1: Energy in the Workplace

The provision of energy in a typical workplace (say, e.g., a mechanical workshop or a confection house) has undergone major change since the beginning of industrialization. We all remember from our youth the time when lathes and milling machines (or weaving stations) were driven from a central shaft that was running through the whole workspace over the heads of the workers, who would connect by sliding a driving flap over a running wheel. When electrical motors became common and cheap, each machine would have one or more individual motors driving it (Fig. 2.1).

2.1.2.2 Example 2: Cars and Public Transportations

We know of two major ways of providing transportation: a road infrastructure for individual cars and trucks on the one hand and, on the other, a system of railway transportation (we also know of mixed systems, public busses etc…). Each has its own possibilities and problems: we'll discuss these a bit further.

Fig. 2.1 Centralized energy provision in an ancient textile mill vs. the decentralized way (Pictures: Wikipedia and ecvv.com)

2.1.2.3 Example 3: The Stock Market

The systems just mentioned are pretty obvious, but our lives are governed by systems whose existence may not be that visible. The stock market, for example, allows people to invest in companies and profit from their investments. On the other hand, the system invites peculiar behavior of agencies that seek profit by speculation and may destabilize the system because of its lack of central control. However, central control can be a major problem as well: we all know that centrally steered economies are not capable of providing sufficient performance to satisfy people's needs.

2.1.2.4 Example 4: Collapsing Societies

In his famous book *Collapse*, Jared Diamond (Diamond, *Collapse*, 2005) gives a number of very instructive examples showing the destructive effects both decentralized and centralized systems may have on society. The book provides a very good motivation to study the issue of centralized vs. decentralized systems and their interaction in more detail.

2.1.2.5 Example 5: The Human Neural System

We may forget that our own neural system provides an outstanding example of a very complex system, in which an equilibrium is achieved between centralized and decentralized control, contributing very much to our survival as a species. Actually, the control happens in many layers (as is very well known from the visual system): it is a "hierarchical system" that ranges all the way from a variety of sensors (tactile, auditory, visual, olfactory, gustatory, neural) to the brains and back to a variety of actuators (e.g., muscles), with many short-cut loops and autonomous reactions. The stability of such a massive distributed system is remarkable and is due to the very large number of feedback connections and inhibitions between the various subsystems.

2.1.3 Properties of Local Versus Global

It should be clear from the examples that the properties of centralized and decentralized systems differ considerably. Here is a short table summarizing the main differences (possibly in order of importance):

Centralized	Decentralized
Works on global data	Scales to individual users
Control is relatively easy	Flexible, adaptive but potentially erratic
Forces specific behavior	Allows for creativity
Enforces uniformity	Stimulates diversity
Inflexible, stiff	Malleable
Often economical	Can be wasteful
Optimizes global conditions	Optimizes local conditions
...	...

Remarkably, both centralized and decentralized systems can produce chaos, although the type of chaos generated by one or the other may be very different, even in similar circumstances. A good, if not so beautiful example is given by our present day mobility and transportation systems. In a public transportation system, the travel time from point A to point B is very dependent on the precise location of both points. From Brussels' Midi station to Paris' North station, travel time can be as little as 1 h and 25 min by TGV, the time many people who live in the periphery of Brussels and use public transportation need to travel to their work spot in the city. This is an example of *chaos in space*. If, on the other hand, you travel by car in Belgium, getting from A to B may depend very much on the time of the day, between Leuven and Brussels, 25 km apart, it may range from 2 h to 20 min. This is then an example of *chaos in time*.

Chaos is an important notion in dynamical system theory, let us now make the notions of *system* and *chaos* a bit more precise. A *system* is commonly thought as being an *assembly of interacting components, with a distinct identity for the external world*. A *dynamical system* is then a system whose state evolves in time, the *state* of the system being the collection of characteristic, in time varying or dynamic, internal parameters (a system often has many characteristic parameters, a distinction is usually made between static and dynamic parameters.). At some point in time, the system starts from an *initial state*, which then evolves according to the *systems dynamics*, which are the physical laws (often differential equations) that give the direction of evolution in function of the actual state and the active inputs (i.e., the interactions with the external world)—for example: the state of our planetary system could be given in a simplified model by the position and velocity of the centers of gravity of the sun and all the planets, and the evolution by the laws of gravity between them.

The evolution is said to be *chaotic*, if small variations in the initial state produce very large variations in outcome after some time, even when the system's state remains bounded. This type of behavior is only possible in non-linear systems, i.e., systems with non-linear dynamics—such as the planetary system just mentioned. One distinguishes chaotic behavior from unstable behavior: a system is *unstable* when its evolution is unbounded and divergent, but evolves regularly. In contrast, chaos typically occurs in a system that does not become unbounded, but whose behavior is extremely sensitive to changes in initial conditions: its trajectories diverge from each other, although they do not diverge in an absolute way. Many nice games (e.g., GO or Checkers) exploit chaos for fun: a small change in an earlier position may yield enormous consequences later on.

Let us dwell a little longer on chaos as an extremely important potential property of non-linear dynamical systems. In global terms, chaos is what makes unpredictability possible, and hence is a necessary condition for what we could call *freedom*, the possibility of making choices freely and going different ways according to whims or insights. How freedom is at all possible, given the rigidity of the laws of nature as well as our human psychological tendencies is a topic of independent interest that we cannot consider here, we simply have to accept that chaos permeates life, and then have to accept the consequences as well.

One extra word on the philosophical issue: is freedom is at all possible? I do not question the existence of short-term determinism of our immediate reactions, but we do have longer term freedom in the choice of further courses of action, after time for reflection. This allows us to set up scenarios on which new decisions can be based, neither of which are predictable. Bach's "Wohltemperiertes Klavier" cannot be deduced from the standard model of physics!

From the examples above we can infer that chaos is handled by different systems in very different ways, some successful, many unsuccessful. Distributed systems that go havoc are, e.g., road transportation systems, the stock market and forest logging and fisheries. But globally controlled systems can also easily go into a chaotic mode, e.g., energy monopolies, plan economies, warfare (consider our two last world wars!), the NSA and what have you. Add to that: there is "good" chaos and "bad" chaos, the first allows for freedom and higher order modes of behavior, while the other causes havoc. There is even "neutral" chaos, much of nature is by nature chaotic and has to be so: without chaos no life!

So the question becomes: how to develop systems in which existing chaos is used beneficially, but which do not run into havoc: a system design issue. Our topic of interest here is that we want systems to be sustainable—lack of sustainability is instability that has to be avoided at all price: it makes the system collapse, often in very harmful ways (a matter of life and death!).

2.1.4 The Design Issue

Designing a (sustainable) system requires the solution of a number of issues, which I wish to consider in sequence, although they are strongly related. Here they are:

- How to reconcile centralized with decentralized control
- How to "beat chaos" (we shall see: by the introduction of intelligence in the system, but that has to be carefully motivated)
- How to optimize system behavior

The end result then should be a system that reconciles local freedom with global stability, and allows for adaptive evolution: maybe getting the best of the centralized and decentralized solutions.

2.1.5 Reconciling Local with Global

There are some traditional approaches to this problem, which I wish to review briefly:

1. *Allow maximal freedom, regulate excesses if necessary.* I would call this the "Adam Smith" approach. It is based on the mostly unfounded belief that freely competing systems will regulate themselves automatically, provided every agent pursues its own interests. That this is wishful thinking is a consequence of the fact that unregulated systems have a strong tendency either to explode or collapse. Nonetheless, any well-functioning system will have to provide for at least a measure of local independence and freedom, the consequence being that regulation will either have to be internalized locally, or has to be restricted in a limited way by delegation to competent parties. How this can be done is our central design problem, I shall claim that this is precisely what requires the introduction of "intelligent control" (or more bluntly, intelligence) into the system.
2. *Keep public ownership, provide user licenses or concessions.* This is what I would call the "Crown approach": the king (i.e., the state) possesses everything, but allows the citizens to make use of the possessions, be it under strong centrally decreed requirements. One may think that such state centralism is characteristic of communist governments, but that is not so. The crown approach was the common modus operandi in the British Empire in colonial times, and forms the basis of much of the legislation in democratic countries (including the U.S.A.). A symptom of it is property taxes: these are only justified if the state considers itself proprietor and the citizens users of the property. A crown approach is necessary to handle common resources such as air, water, clouds, underground, stellar space and what have you. A problem with the crown approach is the concentration of intelligence in a central agency. This prevents adaptability (see e.g., the destruction of natural habitats in Canada and Montana).
3. *Subdivide actors in local,, interacting entities, each endowed with private rights and obligations.* I would call this the "network approach". A good example is the way many countries are handling energy distribution nowadays: a division between producers, distributors and users, all properly regulated. This approach combines some of the properties of the first and second types just mentioned. A step further (continuing with the example) is the introduction of the *smart grid.* Besides a logical distribution of tasks between various actors, the smart grid introduces freedom of action and intelligence, turning the actors into *intelligent agents.* I shall discuss this case in more detail further on.

I now want to show that the way out of the traditional approaches is very much conditioned by the introduction or the allowance of intelligence with the agents that constitute the system. The issue of defining and distributing intelligence proves to be the key to advanced system design.

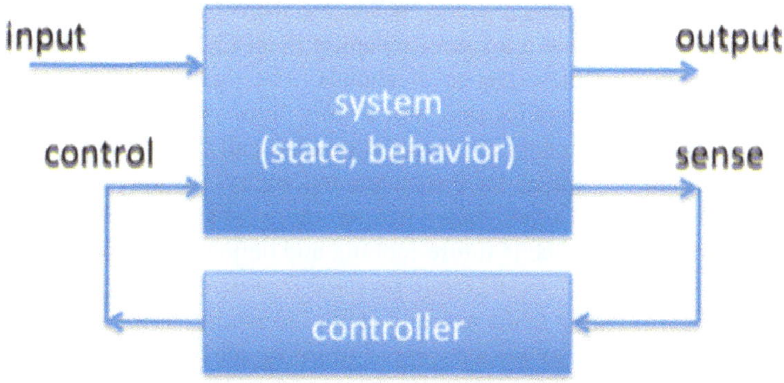

Fig. 2.2 A block diagram describing the control situation. *Blocks* represent system functions, while *arrows* denote connections by communication of signals

2.1.6 Bringing Intelligence into the Control Loop

Control means feedback: the system measures some parameters, uses that information to assess how much the system is deviating from the desired course (it always will, even if a little bit), and computes a correction on the system controls (inputs) to steer the system back to course. Engineers are used to depict such a situation with a block diagram—see Fig. 2.2 above: blocks represent *system functions* and connections *signals* between them. A block diagram is an external view of the system: how it looks from the outside as far as its connections are concerned.

An internal view of the system distinguishes the *system's state*, i.e., all the relevant data that characterize the system at a given point in time (e.g., position, speed, concentrations, temperature,...) and the *system's dynamics*, i.e., all the information needed to describe the evolution of the system, given its state and the inputs applied to it. One can hardly overemphasize the importance of adequate *system modeling*, i.e., the characterization of relevant system's communication channels, state and dynamics (The difference between Archimedean and Newtonian mechanics is in the first place a different choice of state variables (velocity *and* position instead of just position.).

Let us now bring intelligence into this model. The controller itself may be considered to be a (sub-)system as well. It may be autonomous and pre-programmed (as in many automatic systems) or it may use additional resources, such as access to a source of knowledge and the capacity to reason on its findings. *The capacity to produce scenarios, evaluate them and then decide on the next things to do is a central characteristic of intelligence*. It is worthwhile to go a little deeper on this point and to make the connection with "useful chaos" mentioned earlier.

The mechanism that makes intelligence possible is *the ability to give meaning to what is observed or communicated*. Computer scientists make a fundamental distinction between *syntax* or structure and *semantics*, the latter being the *meaning* that has to be attached to a given structure. E.g., a sound has no meaning in itself, but our minds may recognize it as the "u:" in the word "do". Also the sound "do" has no meaning in itself but it acquires meaning in an English sentence, which we interpret as a meaningful sequence of sounds. The duality syntax-semantics is fundamental because semantics does not follow from syntax. It necessitates the incidence of external elements in a process that we call "interpretation". Philosophers and computer scientists have debated at length what they should properly understand by semantics (a field called Epistemology), I suffice here to mention its two main characteristics: one is future-looking and defines semantics as *the consequences a given state of the world may produce in possible future worlds derived from it*, the other is looking to the past and defines semantics as *the assembly of past situations that have led to the concerned state of the world.* Both viewpoints (forward and backward) appear to be necessary for correct 'understanding' and are connected as well: semantics consists of *peering into potential future scenarios while using stored experience obtained by learning or by consultation of external sources of knowledge.*

Although the introduction of a semantic layer on top of syntactical structures is fundamental, the construction of an intelligent system does not stop there. To be effective, semantics has to be formalized and hence turns into a novel structure, which then again is in need of its own semantics, which has to be formalized etc... Sounds become words, words become sentences, sentences become paragraphs, paragraphs become books, books sublimate to theories (e.g., on law), laws are being designed on the basis of legal principles etc... These sublimated or abstracted semantic structures appear on top of random looking "physical" structures—Ilya Prigogine has called this common phenomenon "Chaos but no time to get lost", see (Prigogine and Stengers 1984).

To build intelligence in a controller, one then has to provide the controller with access to outside knowledge (a *semantic context)* and allow it to reason on what it finds, i.e., to develop potential future scenarios and then force the system to make a choice by setting adequate parameters in the controller. This is shown schematically in Fig. 2.3.

A given intelligent controller can obtain its information from access to external knowledge bases (via communication), via a learning method based on past experience, or hard programmed, e.g., as a set of rules (called rule-based control). However, the way it behaves (the controller's dynamics) will be *conditioned* by a number of factors: the way it is programmed, of course, but also by a variety of environmental influences. The way our human brains function illuminates this point: they are influenced by a variety of chemicals, called neuro-transmitters, that translate sense induced emotions into how brains react (e.g., adrenaline for anger, dopamine for pleasure etc...).

It is now not hard to imagine how a distributed intelligent system is put together from interacting individual intelligent systems, Fig. 2.4 shows such a network structure symbolically.

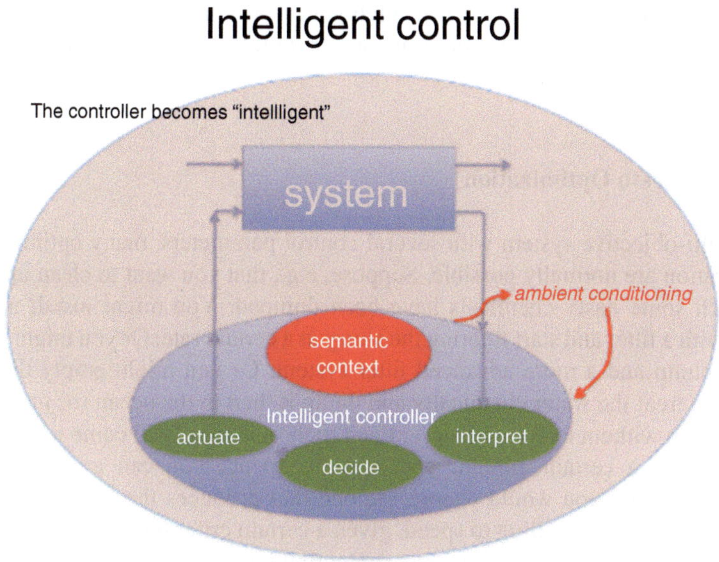

Fig. 2.3 Intelligence introduced in the control loop

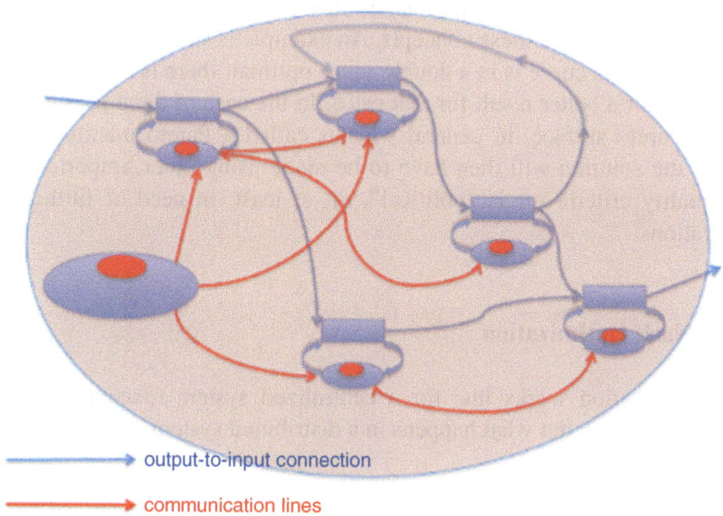

Fig. 2.4 Intelligent systems connected in a network

2.1.7 Optimizing System Behavior

Our model so far allows for a great variety of systems, ranging from fully centralized, where there would be only one intelligent node steering all the others, each normally equipped with its own automatic controller, to fully distributed systems in which each subsystem has its own intelligence on board. Before going into the details of how such systems could operate favorably, let us first explore existing optimization strategies.

2.1.7.1 Pareto Optimization

In a multi-objective system with several control parameters, many optimal modes of operation are normally possible. Suppose, e.g., that you want to clean up a pond in which some nasty chemicals have been dumped. You might install a simple pump with a filter and start filtering the slurry at a certain rate. Or you might install a bigger pump and a more advanced filter system. Or you might empty the whole reservoir, treat the water chemically and dump it then in the ocean (or just dump it in the ocean without treating it) etc… Each such "solution" will come with a certain price. Given a certain amount you want to spend, you can evaluate various possibilities, and you would choose the one that produces the best result for the price you are able or willing to spend, given a certain criterion for "best" (what you consider best will be your context…) Continuing your careful study of quality vs. price, you shall come up with a diagram that shows the best solution given a price, or, alternatively, the best price given a certain result. That would produce a two-dimensional Pareto diagram, named after the famous economist Pareto, who probably first developed these concepts. An example is shown in the Fig. 2.5. Each point on the Pareto curve is in a double sense optimal: there is no better price for a given result nor a better result for that price. In the case of three parameters, there will be a Pareto surface, in general what is called a Pareto manifold. The final choice of the solution will then have to be made using other, imported or conditioned quality criteria—it is "political", or, at least, in need of further external considerations.

2.1.7.2 Nash Optimization

Pareto optimization works fine for a centralized system (except when there is chaos—see further), but what happens in a distributed system of intelligent agents?

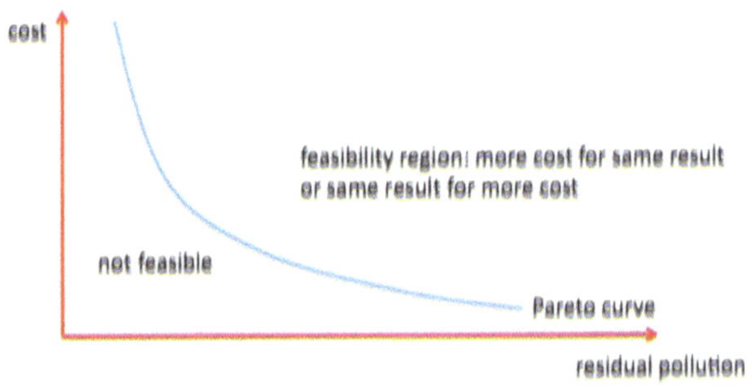

Fig. 2.5 Illustration of Pareto optimization

Table 2.1 The prisoner's dilemma

Gain A/Gain B	A betrays B	A remains silent
B betrays A	20 years prison/20 years prison	30 years prison/free
B remains silent	Free/30 years prison	10 years prison/10 years prison

Nash (1951) has studied this issue, assuming that each agent has to make a decision independently, without being influenced by the decisions of the agent's peers (e.g., because of lack of knowledge, as would happen in a game of chess). Nash constructed beautiful examples that show that the individual decision making is far from optimal, even sometimes the worst possible. The most famous example is maybe the "prisoner's dilemma", where prisoner's would rather betray each other than keep their mouth shut, each expecting better treatment doing so at the cost of the common good. Breaking the prisoner's dilemma (and all other examples of bad Nash optimal points) always involves improving communication between the agents, or else, if such is not possible, improving the quality of the conditioning environment (Table 2.1).

2.1.7.3 The Case of Chaos

The Pareto manifold is presumably a smooth surface, reflecting a non-chaotic property: close-by criteria have close-by optimal points. However, as we have seen, chaotic situations are very common, especially where organizational problems are concerned. I already gave the travel time example, here is another. There is a lot of discussion about whether CO_2 concentrations in the atmosphere will reach a *tipping point*, at which time the arable lands on earth will massively re-order. Tipping points are a symptom of chaotic behavior, as are *bifurcations*, when the system's evolution splits in different possible branches, as happens on a saddle: one can go down one way or the other. The new course chosen by the system is not necessarily bad, but may require major adjustments, which in turn requires major intelligence from the participants to adapt, hopefully based on a good understanding of the new dynamics.

In many systems there are sequences of tipping points, often leading to a complete breakdown of regular behavior. What is going to give direction to future evolution in such cases? An answer can already be inferred from our previous discussion: one has to move to a new level of understanding, i.e., a new semantic level, using more global intelligence. One strategy is, of course, to try to avoid the occurrence of the tipping point, i.e., influencing the dynamics in such a way that the system is steered away from it, but this requires understanding the dynamics and the use of intelligence just the same. Some optimality can be achieved in a chaotic situation, but only in a statistical way (often there is no clear optimum in the basic system), which anyway requires moving to a higher, more global level at which a statistical analysis is possible. E.g., optimizing travel times in a public transportation system would aim at making as many travelers as possible happy, "happiness" being a condition the original system does not know about.

Intelligence of the agents and communication between them are key to effective and adaptive distributed control. While this may now seem obvious, the problem is of course how to design such a system, thereby optimizing desired properties, such as sustainability and resilience while keeping costs affordable. In the next section I discuss this question, without being able to give full answers, because the field of distributed system design is large and still very much in its infancy. Nonetheless, we already know of some very well designed such systems, the "smart grid" being one of them, which I will give as an example in a following section.

2.1.8 Designing (Distributed) Systems

Centralized control is often advocated on the basis of competence (only the central authority has the necessary competence), or on the basis of necessary stability (only the central authority can guarantee stability of the overall system). The issues of competence and stability are of course genuine, but they do not make centralization necessary. Also intelligence can be distributed, and as we have seen, a distributed system is potentially much more adaptive than a centralized. Nonetheless, there are limits to the possibility of distributing intelligence. E.g., individual users of electrical energy cannot oversee the needs of the overall network. Any somewhat complex system will therefore have some hierarchical structure, whereby some agents have specialized intelligence and more control than others. This brings up the question of how agents with different capacities deal with each other: the communication issue.

Let us go back to our considerations on syntax and semantics. Each communication channel consists of a physical medium over which structured messages (signals) are channeled. The structure of these messages reflect the semantics of the sender—they have meaning for the sender and are purposed as either information for the receiver or as notification of a behavior desired by the sender. On the other hand, the receiver is obtaining the same structural data (maybe somewhat corrupted by noise or interferences), and then starts reinterpreting it according to its own semantics.

In other words: sender and receiver share syntax but differ in semantics. Much of misunderstandings between people or even between countries are due to semantic differences. The same words mean different things to different people; each word carries for each of them historical (and hence contextual) connotations. As a result, for good understanding between parties an *alignment of semantics* is necessary. This does not mean that the semantics of both parties has to be the same, but it has to be aligned, in the sense that the sending party can have confidence in the interpretation of the receiving party. Unfortunately, the semantic alignment that has to be developed in many modern very complex systems is highly technical. Mastering the total technical situation is next to impossible for any individual agent, and as a consequence, individual agents will have to make their individual expertise shareable to the community, while the community, in return, will have to resort to *trust* of agents with specialized expertise. Are there mechanisms for this?

Trust is a question of correct estimation by one agent of the behavior of another, given shared intelligence. Or: why would an agent act according to the wishes of another? It is a question of motivation by conditioning! It is often thought (mostly by economists) that all motivation of an intelligent (or even a super intelligent) agent is the maximization of individual profit—the Adam Smith principle. There is even a belief among economists that leaving all to maximization of individual profit will stabilize the system automatically, because if all agents behave in this way, they will equilibrate the system (never mind if it is a multi-dimensional saddle). The implicit assumption in such a model is that all agents are equally expert, that communication is universally effective and that there are no semantic problems between them—potentially the most utopian world.

Without diminishing the importance of profit, let me mention other conditioning factors that are effective in many situations. Here is a little list:

- Profits, gains, advantages
- Ideology: hatred, nationalism, beliefs
- Generalized opinions, examples of others, herd-spirit
- Learning, education
- Pleasure

Each of these may be active, even in combination. E.g., whether one expects to make profit from some actions may be dependent on perceived experience of others, or on presumed knowledge etc.... Some of these factors can be influenced in positive but also in negative ways. Knowledge is always partial and may be manipulated, there are conflicts of interests, power struggles, stray conditionings etc.... It is for sure important that the sharing of intelligence is done in as an objective (i.e., non-manipulatory) way as possible, although even that is no guarantee of correctness. We live in an imperfect world and have to conceive our systems in a way that make them as trustworthy, i.e., behaviorally predictable, as possible, given the uncertainty (a property called *robustness*). Important is that these aspects are taken into account when designing a system.

Let us recount some strategies to achieve trustworthy system behavior.

Strategy #1: enforce scenarios by reward and punishment (the classical authoritarian way). A modern example is the definition of pricing policies for distributed energy production/consumption by a central authority such as the network manager.

Strategy #2: share potential scenarios, and elect one that is viewed as the best one by a majority of participants (the democratic way).

Strategy #3: make scenarios developed by individual agents compatible through the creation of win-win situations (an inclusive approach).

Needless to say, in all practical situations a combination of these three strategies will be necessary. While strategy #1 is by far the simplest and most effective, strategy #2 will be the fairest and strategy #3 the most motivational, but these two latter strategies are more cumbersome and time-consuming than the first and not always possible or effective.

The present way of thinking then leads to the following design considerations:

1. A distributed system with distributed control needs sharing of information (communication) and *alignment of semantics* (mutual understanding) between acting agents, if anything like close to optimal operation is to be achieved; one may summarize this aspect as *creating a "win-win" environment*
2. In a complex intelligent system not all agents are equal as far as expertise and power to act is concerned (see also the next point on network stability); given the necessity of creating win-win situations among all participants, the technical expertise of an expert agent has to be abstracted to the level of expertise (i.e., the semantics) and power of other participants individually or at least, per class; this can be made effective through mutually acceptable conditioning, either profit sharing and/or various mutual benefits
3. The stability of a distributed system is an issue of central concern, including such a long-term stability issue like sustainability; all participants have to contribute to it in a way that benefit is experienced by all; this is an especially critical issue as the taking of short term benefits by the most aggressive participants may compromise longer term stability. This brings up the issue of "system ethics", which can only be centrally agreed upon, organized and enforced. I shall consider this topic in the next section
4. The alignment of intelligence, necessary for favorable and adaptive development, requires sharing of the assessment of future scenarios, alignment (or maybe even consensus) on strategy, communication and feedback concerning effects, benefits and losses, access to global information and, last but not least, access to decision making and the exercise of power at an adequate level of influence. This is the true meaning of democracy: having an adequate share in the exercise of power, based on access to the best possible knowledge
5. In a distributed environment, decision making parameters are influenced (regulated) through conditioning; although cost or profit will often be the main conditioning factor—and win-win situations will have to accommodate shared profit—other factors may play an important role as well. Long term benefit may clash with short term profit taking, intelligent quality assessment may mitigate the pure strive for profit, and ideological considerations may strongly influence the decision making, including herd behavior (such as 'all my friends chose for the more expensive option'...)

A distributed intelligent system is far from predictable, but it contains the elements needed for an evolution that can correct itself, thereby ensuring stability, and steering the community to a more desirable future, taking into account increasing experience and acquisition of knowledge.

Nonetheless, a number of structural elements need to be present to allow proper functioning, be they "democratically" sanctioned. In particular: (1) global ethical agreement between participants is needed to keep corrupting interference by the various agents in check (see the next section), (2) structuring according to expertise and the ability to exercise power is needed for effectiveness, and (3) proper conditioning has to permeate the system, insuring that its components will act reasonably at any point in time (at least in an acceptable statistical way: one may tolerate deviations as long as they do not threaten the stability of the system as a whole—the property that ensures this is called *robustness*).

2.1.9 A Potpourri of Illustrative Examples

- Avoiding lead in gasoline: ethics set by authority after public sharing of intelligence concerning health; positive acquiescence by all players
- Drug suppression in the United States (and many other places): ethics set, not by consensus but by authority, sharing of intelligence between players woefully lacking
- Reducing CO_2 in the atmosphere: after promising start with emission rights, poor implementation of the win-win situation. Lack of common intelligence between players
- The German "Energiewende": politically driven ethics with strong enticements towards a win-win situation for all participants; special is the strong motivation towards sustainable technology as component of the enticement
- Ocean pollution: a disaster so far: no common ethical understanding
- Fluoro-carbons in the atmosphere: a strong case of intelligence sharing and enforced, publicly supported ethics

2.1.10 Ethics

In the Western world, the term "ethics" has often been confused with "morals". It is time to put the matter straight, in view of its importance to understanding the functioning of a distributed system as a whole. I follow in this the treatment of the late Bernard William (Williams: Ethics and the limits of Philosophy 1985). To simplify matters, let me suffice with a working definition: *ethics is the explicit or implicit recipe of behavior adopted by an agent in its dealing with a system.*

The definition invites some comments and clarifications:

1. Ethics has to do with *actual behavior*, not with *what should be done according to some set of behavioral rules*. It just permeates life as it is. One may understand somebody's ethics by observing her/his behavior. It is the accumulation of habit, learning and experience that has been put in place over time to condition the actual way an agent will behave in a certain situation. A medical doctor, an engineer and many other professionals have learned to act in a certain way given a certain situation—that is the ethics of their profession. In many cases this does

not involve any morals: another way of behaving may be just as good (e.g., a medical doctor may prescribe antibiotics for a bad cold and another not, all with good grounds.).

2. Central in the functioning of ethics is an explicit or implicit assessment of *quality*. Necessarily, this will depend on an (again implicit or explicit) method of evaluating quality. This method (I could call it an evaluation system) forms an active super-layer of the system. It is, in a sense, an extra semantic layer: the agent's actions get meaning in the light of his/her ethics.
3. The existence of an ethical layer is unavoidable: even random actions have "randomness" as ethical principle—if this is the case in an actual system, it is an important and potentially dangerous characteristic of it (consider e.g., the stock market). "Laissez faire, laissez aller" can be good but also very bad, because it influences the system's main properties such as stability or sustainability.
4. The role of Morals is much more limited: it involves *obligatory behavior (in particular under critical circumstances).* Many professions have Codes of Conduct for professional behavior and critical cases. No doubt, Morals are part of Ethics, but only a relatively small part. Most of our actions take place in non-critical circumstances, may not be essential and hence need no moral intervention. The ambient ethics will actually determine what is considered an essential behavior or a critical circumstance. Critical professions such as Medicine have what is called a "Code of Ethics"—which could better be called a "Code of Morals"—defining what is essential behavior and what is a critical situation, and how a practitioner should deal with those. Such codes can evolve with time, in particular with the increase of expertise and/or knowledge, none of which is absolute. Even such an absolute moral principle as "thou shallt not kill" is hardly helpful in modern medicine (another problem is involved here, namely that an abstract principle needs to be translated to a concrete situation, again a non-trivial ethical step.).
5. In a distributed system, an overall "system ethics" will be active. This is often a matter of life and death—we all make decisions all the time that will secure our permanent well being! In a biological system, the actual ethics has evolved with time and has been sanctioned by evolution. To understand the functioning of a man-made, artificial system, there is no way out except making its ethics explicit. Again, ethics is not cut in stone and in need of evolution, so the system has to provide mechanisms for doing so, which in turn brings up all the system design issues such as stability, resilience etc...

The implementation of pervasive, desirable ethics in a distributed environment (even in a centralized environment) is not an easy matter. Luckily, when the seeds are well sown, the system is allowed to evolve, to acquire knowledge and to adapt on ever changing circumstances, most notably the growth of the system and the addition of many components. As all actions have effect on the whole of the system (or, maybe, only on cloistered parts), the current ethics of each individual agent may have global consequences, especially when it generates conditioning factors.

Indeed, the implementation of "desirable ethics" requires a whole amount of feedbacks: agents will generate ethical requirements on other agents, at several levels of endeavor. At the global level, the issue will be agreement on the ground principles, at the level of local implementation, the building up of trust between agents through communication and sharing of intelligence and at the level of activities, oversight and conditioning.

Since the ethics sets up both global and derived local behavioral conditioning, it will not go without constraints. E.g., when parties have made an agreement, its faithful execution requires constraints on the contractors, and hence sanctions if they desist. Again, and keeping in line with our principle of shared intelligence, the mechanism of sanctions has to be mutually agreed on, has to be reasonable and has to respect individual empowerment.

Much of the ethical layer becomes "hard coded" in its participants. It takes typically a long time to build it up, because it needs the accumulation of experience and consistency in actual decision-making. It puts the parameters in place (often in the form of a rule base) that will allow the agents to act instantaneously and hence automatically. Ethics lives at a very different time scale than the scale at which the system operates, and necessarily so (think of driving a car: before you can do that in a stable way, you had to consolidate all your driving habits).

This is not to say that there is no environmental influence on how the system operates: conditioning factors do play a distinct role in the instantaneous parameter setting (e.g., a person reacts differently when angry than when composed), but even their influence is set previously by the intelligent, ethical layer (e.g., some people will always suppress their anger.).

Designing systems with their ethics in mind is a pretty recent phenomenon in engineering. Most of past engineering was geared towards achieving specific results, often using novel technical means to do so. This bottom-up approach contrasts with the top-down system construction needed in our modern, highly technology determined societies. In the bottom-up approach, individual local systems will be equipped with local feedback loops ensuring the operational stability of these subsystems. This is of course a first prerequisite for overall stability. However, the needed top-down system design would traditionally be restricted to a centralized hierarchy, in which the only upward information would consist of a limited set of major performance factors, often independently acquired, followed by purely top-down control of available parameters: hence no sharing of intelligence! A well functioning distributed system needs a much more elaborate network of feedbacks, both "upwards" and "downwards", so elaborate that the distinction between "up" and "down" disappears, to be replaced by a strongly connected *network* of interacting intelligent agents.

Although stability may seem to become a major issue in such a network, the situation is saved by having the decision making organized in distinct semantic layers, each responsible for a well defined and controllable environment. E.g., the ethics layer, which includes system definition principles, lives in a totally different world than any operations layer—totally different both in content and in timing. The interaction between these layers will have different timing properties: control is executed and stability insured at very different time scales. The connection and sharing of information (feedback) is done by components in each agent's intelligence layer, which exchanges and interprets information provided and/or acquired by sensing or observation. In many natural biological systems, long term stability is assured by apoptosis of short term acting agents, a comforting thought for sure.

2.1.11 Desirable and Actual Ethics

The ethics in the mind of the designers of a system is not necessarily, even most often than not, the same as the actual ethics: the road to hell is paved with good intentions. One may be sorry over this state of affairs, but one can better be realistic as a system designer. We already talked about "alignment of semantics" to achieve effective communication between agents, and "alignment of intelligence" to enable the sharing of scenarios and decision making process, here we encounter the next level of necessary alignment: "alignment of ethics", to ensure that the system is so designed that actual ethics and desired (or designed) ethics agree with each other. There are a number of means available to designers to achieve this. The conditioning factors play an important role because they are active at the global system level, just like ethics. One main principle is of course that it should generally be more advantageous for all agents to behave according to the desirable ethics than otherwise—necessitating the implementation of rewards and punishments, with probably a much higher emphasis on rewards. But all the other conditioning factors are important as well and can be used to great benefit, especially so as to make the system more robust: education, ideology, common opinion building, examples etc… (The effectiveness of a specific conditioning factor is an important design parameter.).

Time to move to an example in which such ideas have been well implemented, and a number of the issues raised can be observed directly.

2.1.12 The Smart Grid as an Example

Given all the design considerations and constraints mentioned so far, one may wonder whether such a thing as a man-made complex intelligent system actually exists. Remarkably, the answer is 'yes', there are actually quite a few of them, but a prime example is the fairly recent introduction of *smart grids* for electrical energy generation and distribution in many countries, including very big ones like major states of the United States and Germany. It is the merit of the American National Institute of Standards and Technology (NIST) to have provided and standardized the main ingredients of its design. But before diving a bit in the details, here is how the European Technical Platform for Electrical Networks of the Future defines it:

> A Smart Grid is an electricity network that can intelligently integrate the behavior and actions of all users connected to it—generators, consumers and those that do both—in order to efficiently deliver sustainable, economic and secure electricity supplies.

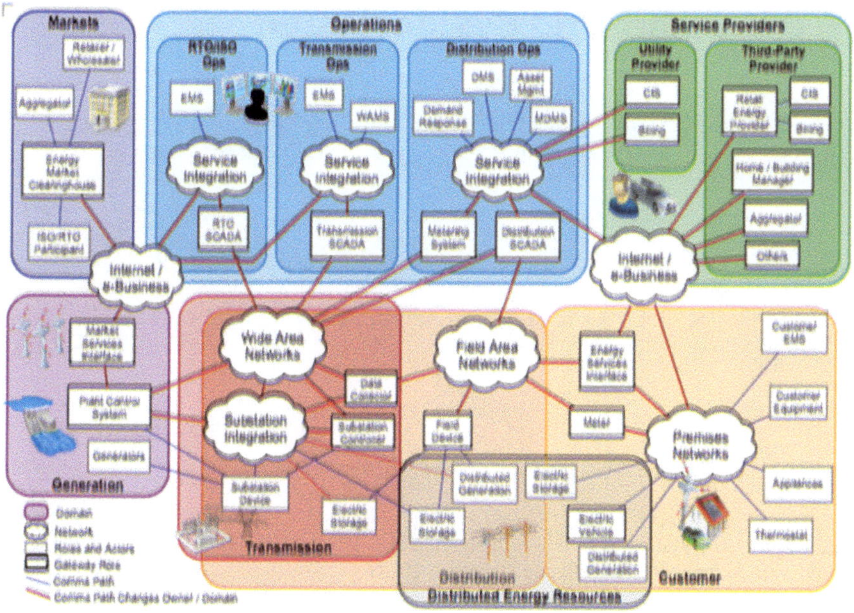

Fig. 2.6 Participants and their relations as described by Arnold, IEEE Proceedings, June 2011

A short survey of the participants and their interconnections in a smart grid according to Arnold, IEEE Proceedings, June 2011, is as shown in Fig. 2.6.

A system view describing the various levels at which participants in a smart grid operate is given in Fig. 2.7, taken from a presentation by Prof. Jacobsen of the TUM Middleware Systems Research Group.

A top-level smart grid reference model as defined by NIST, shown in Fig. 2.8.

Although it would be inappropriate to start describing the complete NIST system here (there is ample information on the NIST website and in the literature), let me summarize some of its salient ingredients:

1. The system consists of a network of very diverse agents, with very diverse sizes and interests; there is, however, a central agent that operates the network and is responsible for its effectiveness and stability. This agent derives its profit from precisely these tasks;
2. The network offers to all participants flexibility and possibilities (i.e., freedom) they did not have before in a pure supplier-consumer model; in particular: it allows for "pro-sumers" i.e., parties that generate as well as consume electricity, for new forms of energy generation and storage (i.e., for adaptation) and for various forms of providing supply and consumption;
3. Although all the participants have very different abilities and empowerment, there is considerable alignment of intelligence in the system. To achieve a decent win-win situation among all participants, it is important that these semantic and ethical aspects are continuously managed by an entity that has both the power and the motivation to do so.

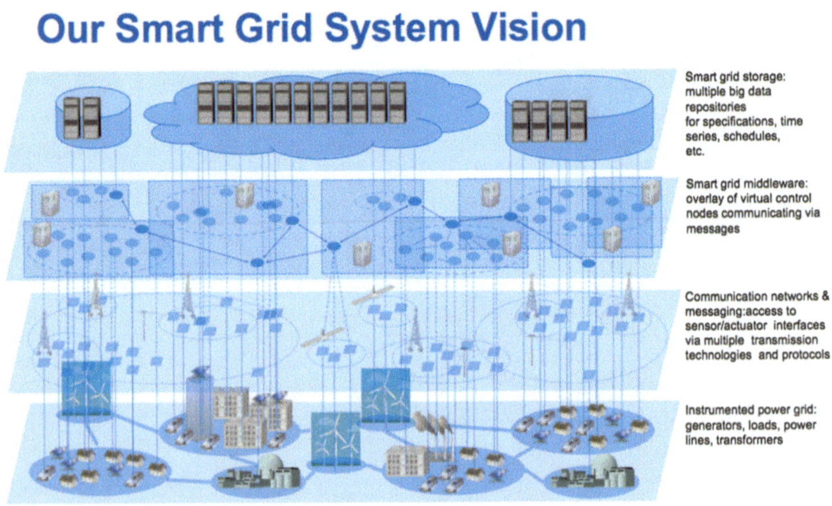

Fig. 2.7 Layered system overview of a smart grid, as defined by Prof. Jacobsen of the TUM Middleware Systems Research Group (Jacobsen 2013)

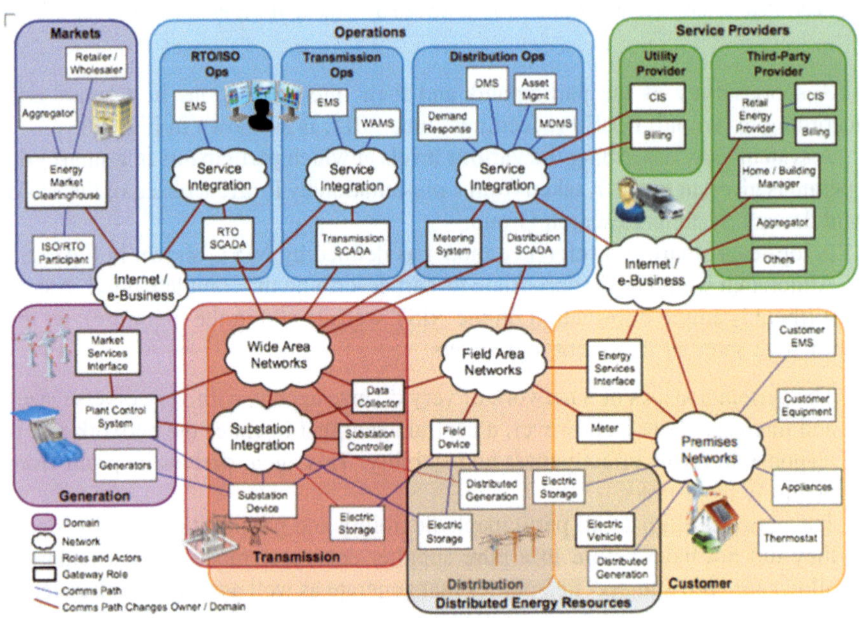

Fig. 2.8 Smart grid reference model as defined by NIST (2014)

4. The system exhibits several control levels, actually each participant has one. These control loops differ considerably from each other dependent on the type of participant. A large power plant must insure the stability and synchronization of its AC-production capacity, which is critical due to the inertia of its generation equipment (the main source of instability in an AC power network) but locally achievable. The network provider, on the other hand, must insure the stability of the voltage level and the frequency of the AC current it supplies. It has the complex task of equilibrating supply and demand, while taking the needs of both types of participants into account. The pro-sumers and con-sumers are typically small-scale participants, but there is a large number of them, so control can be largely statistical, using appropriate conditioners. Their flexibility and the power of modern electronics allow them to play the central role in stabilizing the net. All these distributed controllers have to align themselves by exchanging information and intelligence, the next point I want to consider.
5. On top of all these controllers, there is a layer of intelligence alignment, made possible by instantaneous communication over a distributed communication network (the internet). The proper functioning of the network is of course crucially important, but even more important is how the semantic alignment is made, or to put it more colloquially: how parties understand each other. I hereby entice my readers to explore this matter in detail, because the smart grid does a very good job at it: it defines the scope of interest of each participant, both in terms of structure and semantics, what kind of information has to be exchanged, how the mutual conditioning is achieved (the win-win situation), how mutual expectations can be fulfilled, how security can be achieved and what the available freedom is for each.
6. And what about ethics? It should be clear from our previous discussion, that the principles used in defining the sharing of intelligence, the control scope, mutual benefits and obligations (i.e., the win-win situation), in one word: the system design principles, amount to its ethics and, as mentioned before, these may be "conscious" or "unconscious". To be more precise on this point: when a smart grid system as described is implemented in a country (like it is in many countries already), this amounts to an ethical choice, sanctioned by the hopefully democratic establishment of that country. With ethics as normal behavior also come "morals" as a tiny but necessary part of it: how participants have to behave in critical situations or to keep the system functioning. In the case of a smart grid, a whole emergency system has to be in place that will force participants to behave in certain ways when, say, the network stability is at stake. But morals will not only be involved in structurally critical cases. Also the honest interaction between participants will require enforcement, so that unwieldy profit taking of certain participants using their immediate power at the cost of others or even at the cost of the entire system is duly sanctioned or prohibited. In addition, the ethical set up is in need of evolution and adaptation depending on assessment, learning and increasing knowledge. Part of the ethical layer is then how such evolution is organized and maintained. Different systems may be adopted for this, this issue would need a separate treatment beyond the scope of the present paper.

And the result? The smart grid is an eminent example of the right combination of distributed and centralized control. It achieves excellent performance on all system quality criteria:

- *Stability:* thanks to clever partitioning of tasks, sharing of intelligence and the clear definition of the role of each participant, stability is achieved at all levels;
- *Sustainability*: the system is able to offer the necessary enticements to assure that all participants strive to maximal sustainability;
- *Resilience:* in case of emergency or failure of some components, the system can easily adapt itself to a new stable operating point; this is mainly due to the increasing abilities of modern electronics (a smart grid is more stable and resilient than a traditional grid, because the components that can cause instability are better shielded from the overall system, and the other components can adapt easily to new situations. This is remarkable because the opposite was predicted some 10 years ago—wrongly as it turned out.)
- *Robustness:* can the system tolerate behavioral deviations from the nominal design criteria by some or a large collection of its agents? This is an important question that is not easy to answer, if only because so many behavioral variations in different directions by different agents would have to be taken into account. Part of the solution adopted in the smart grid, is by careful monitoring of the behavior of all the agents concerning the main characteristics of the system (not only its operating points, but also the use of resources, profit taking, investments, use patterns etc…) That then reduces the problem to the robustness of the network operator, a question that can only be dealt with at a higher level, in this case probably a political level. Nonetheless, it remains important that in the design of a specific instance, sufficient leeway is build in, so that the network operator does have a sufficiently broad feasibility range, plus the ability to continuously monitor and enforce proper operation;
- *Adaptivity:* it should be clear that a smart grid scores optimal for this criterion!
- *Optimality:* in the beginning of this paper I hinted at a potential problem of wasteful use of resources in a distributed system. What is the case here? The concentration of electrical power generation in large fossil burning or nuclear power plants certainly provides for the best possible use of resources *in that kind of technology*. However, a change of technology changes the picture as well. A new technology may offer advantages that the previous did not have. Although it is more efficient to concentrate solar cells in larger photo-voltaic (PV-)generation plants, local power generation with PV allows for a much more flexible use of resources and local independence. This will even increase in the future, as the performance of PV is currently increasing very fast for little cost, making PV in the long run feasible even in countries with less solar influx. At some point, flexibility wins over efficiency, and this will predictably be so with PV (for wind power the verdict is still out!)

Taking everything together, the smart grid is a quantum step improvement on existing systems and can be used as an excellent example of advanced distributed system design, for which the American NIST deserves a lot of credit (I understand the EU is adopting its main principles as well!).

2.1.13 *Final Remarks*

The analysis in this paper has been mostly qualitative. However, many of the notions and properties that were covered can be made quantitatively precise: individual agents can be numerically characterized (including control loops), their "intelligence" precisely described, communication links can be made explicit (including the necessary "alignment of semantics") etc... For some central agents (e.g., the network operator in a smart grid) such descriptions may be very elaborate, and a whole array of technical principles may have to be brought in line. However, the important point is that for major systems, such an approach *is feasible*, if not necessarily easy. The 'smart grid' was given as an example, but there are many other successful examples, as well as many attempts at defining new systems, in particular environmental systems (a recent example is "Bio-economics"). All the issues discussed in the paper come to bear, often in a critical way, and in particular, 'actual ethics' as defined plays a decisive role, which system designers have to take into account if they want to be successful at all. Many well-intentioned systems have failed on this point (Bio-economics being a recent example of failure at the ethical level.) Let this paper be an invitation to careful analysis. Hopefully, it has provided some of the key notions and issues to be considered.

2.2 The Principles of Subsidiarity and Internality as Ordering Principles of Decentralized Structures, Illustrated by the Example of Water Management

Ulrich Drost and Martin Grambow

2.2.1 *Key Messages*

Decentralization presupposes an ordering principle to distribute tasks between a central organization and decentralized structures. The principle of subsidiarity has been developed over the course of thousands of years for state structures in the interest of distributing tasks in decentralized hierarchical structures.

The management of water resources in river basin districts in European water management has led to a new approach in the decentralization of tasks outside of traditional subsidiarity structures. These structures have in common that the decentralized units have to approach their tasks with internality, i.e. taking responsibility and based on their own merit, in order for decentralization to be successful.

Dealing with water as a global common has always been subject to tension between centralized and decentralized/subsidiary management. This paper will lay out and discuss the manner in which water as a global common is addressed as example for building decentralized structures on international, European, national, regional, local and individual levels. The need for internality to attain successful decentralized structures will be demonstrated.

2.2.2 *Subsidiarity and Internality within the Scope of Future Development of Decentralized Structures*

2.2.2.1 The Principles of Subsidiarity and Internality as a General Approach

The ambivalence between centralized and decentralized structures appears to be a significant component of state-building. Subsidiarity was first formulated as a general principle during antiquity by Aristotle, in his treatise "Politeía". Accordingly, man is a "political animal" that builds states, forming communities on various organizational levels to fulfill the necessary tasks. According to Aristotle, the state is the maximum organizational level. It is all-encompassing and self-sufficient. However, its responsibilities are limited to aspects that an individual or a domestic or village community cannot attend to.

Since then, the principle of subsidiarity has been developed further as a general guiding principle and structural element in state theory. In the western world at least, subsidiarity has entered into all areas of government and social action as a general principle. Accordingly, smaller efficient units have priority in undertaking action, with superordinate organizations assuming accountability and providing support. From a state organization perspective, subsidiarity should be implemented in a hierarchical structure of superordinate and subordinate regulatory authorities (competences). In this hierarchy of competences, superordinate entities must maintain the overall system functional by defining the scope of action for subordinate actors and steering their actions, i.e. they "control the self-control".[1] To that end, subsidiarity not only means rights, it also translates into on both sides.[2] Subsidiarity is a dynamic principle resulting from constant correction and enhancement. It can push in both directions, i.e. from a superordinate structure to a subordinate one, and vice versa. Under certain conditions, this process is even capable of forming new structures. One example is the water sector, where in addition to state hierarchical structures there are also structures specific to certain river districts currently in formation. These structures are established in the interest of attaining predefined management objectives, and are reliant upon compliance with the requirements for objective attainment in all sub-structures, river basins and sub-basins.

[1] Zippelius (2011).

[2] Vogt (2009).

Internality is the conviction that the results from action tend to be experienced as being self-induced instead of being attributable to external causes.[3] The term originates from psychology and of course cannot be directly applied to objects like state structures or bodies of water. However, internality is important for the order of decentralized/subsidiary structures to the extent that such structures must have the capacity to perform their own tasks internally. Otherwise, externalization would counteract the shifting of tasks pursued in decentralized or subsidiary structuring. The principle of subsidiarity always makes success immanently conditional upon the subsidiary unit's internality in task execution. This necessarily also applies to other decentralized structures.

2.2.2.2 Natural Guidelines in Water Management

Water is indispensable. It is the foundation of all life. It constitutes a global principle of water circulation, constantly renewing itself, feeding bodies of water of all shapes and sizes (surface water, groundwater, seas).

Water is a gravimetric resource: the flow mainly available to humans is determined by gravity. This simple formula translates into an intrinsic, ubiquitous upstream and downstream relationship among all water users, i.e. all life. Ergo, there is potential for universal competition for a single resource. This competition demands social order and management. Thus, water management is defined as a basic civil necessity, and therefore a public service (see Sect. 5.1). Since water management is only carried out for inland water and groundwater, in addition to (random) political demarcation the basic management approach for flowing bodies of water is also comprised of management of river basin districts, river basins and sub-basins.

2.2.3 General Legal Guidelines for Water Management Considering the Principle of Subsidiarity

2.2.3.1 International Provisions

In the past and present alike, the provisions applicable to water management across all "hierarchy levels" from individual right to international law have been influenced by the principle of subsidiarity. National provisions in the water sector have long been considered the sole incarnation of central rules. International

[3] Schuler et al. (2001).

provisions have largely come into existence over the past 150 years on the European level in particular, with examples including the Rhine and the Danube.[4]

International treaties for water management overlap with water management areas, which generally exceed national borders. For example, the Convention for the Protection of the Marine Environment of the North-East Atlantic (OSPAR) of 22 September 1992[5] has the objective of taking all possible steps to prevent and eliminate pollution carried in from inland water and the necessary measures to protect the maritime area against the adverse effects of human activities so as to safeguard human health and to conserve marine ecosystems and, when practicable, restore marine areas which have been adversely affected (Art. 2 (1) (a)).

One particularity within the transnational approaches in international maritime law is the resolution adopted by the General Assembly of the United Nations on 28 June 2010, acknowledging access to safe drinking water and sanitation as a human right that is indispensable for enjoying life and all human rights.[6]

On 30 September 2010, the United Nations Human Rights Council also adopted a resolution[7] affirming the human right to safe drinking water and sanitation (HRWS). In this resolution, the HRWS is derived from the right laid out in the ICESCR[8] to a reasonable standard of living and is linked to the right to health as well as the right to life and human dignity. Thus, it can be said that the human right to safe drinking water and sanitation has been secured by the United Nations since 28 June 2010, and that this right also originates from other rights that are laid down in the UN's ICESCR. The latter is of significant importance for the third-party effect of the HRWS. As part of the human right to an adequate standard of living and health as laid down in Articles 11 and 12 of the UN ICESCR, the human right to safe water and sanitation needs no further convention or international regulation. The ICESCR binds all 160 signatory states directly, with their citizens being the direct beneficiaries.[9] Thus, the signatory states are obliged to exploit all their capabilities to take measures, in particular legislative measures, in order to attain full implementation of the rights recognized in the ICESCR on a national level or via international aid and cooperation, especially of an economic and technical

[4] The Danube Protection Commission of 1856, the Central Commission for Navigation on the Rhine, Mannheim Convention of 31 March 1831.

[5] Germany adopted it into law via the Act of 23 August 1994 regarding the International Convention for the Protection of the Marine Environment of the North-East Atlantic (Federal Law Gazette 1994, p. 1355).

[6] Resolution of UN General Assembly (A/64/L.63/Rev. 1 and Add.1).

[7] Resolution Human Rights and Access to Safe Drinking Water and Sanitation (A/HRC/RES/15/9).

[8] International Covenant on Economic, Social and Cultural Rights (ICESCR) of 16 December 1966. *Resolution* 2200A (XXI) of the General Assembly of the United Nations. Entered into force on 3 January 1976.

[9] See explanatory notes from the German Federal Government in Bundestag printed paper 17/4526 p. 5.

nature.[10] However, the rights yielded here are so called progressive rights, which mean one cannot assume that their implementation will occur from 1 day to the next.

Nevertheless, this procedure represents centralization in the definition of standards within the context of a global approach: states are now more bound than was previously the case to implement the human right to water and sanitation.[11] This also includes protection of water resources. Such protection must be comprehensive in order to provide the corresponding state guarantee. When taking a closer look at the water management correlations necessary for this task, it appears that guaranteeing the human right to safe drinking water and sanitation requires much more than anchoring domestic water management in international law. The HRWS ultimately requires a more comprehensive perspective. To that end, it can be stated that at least significant portions of integrated water resource management (IWRM) are necessary for states to guarantee the HRWS.[12]

From the perspective of the principle of subsidiarity, the HRWS guarantees an individual right held by every natural person on a global level, yet does not define rules for implementation. Thus, the states bound by this obligation have the entire spectrum of organizational options at their disposal for decentralization and implementation based on subsidiarity aspects.

2.2.3.2 Supranational Provisions

While in international regulation multiple centralized structures come to an agreement on a common procedure, supranational conventions create a new center. Such interplay between (central) supranational standards and (sub-central) implementation in nations and regions is clearly created in the EU's setting of environmental standards: the Treaty on the European Union generates supranational law, which applies in the respective Member States. The European Union is characterized by the fact that the Member States have conferred competences to it in order to attain objectives they have in common (see Art. 1 (1) EU Treaty[13]). All competences not conferred upon the Union in the treaties remain with the Member States pursuant to Art. 5 of the EU Treaty (see Art. 4 (1) EU Treaty). Pursuant to Art. 5 (1) of the EU Treaty, the principle of limited conferral governs the limits of the Union's competences. The principles of subsidiarity and proportionality govern the use of the Union's competences. Under the principle of limited conferral, the Union will only become active within the limits of its competences that the Member States have conferred upon it in the treaties in order to attain the objectives laid down therein.

[10] See preliminary remarks in the Annex to Optional Protocol Resolution 2200A (XXI) of the General Assembly of the United Nations. Entered into force on 3 January 1976.

[11] (See BT printed paper 17/4526 p. 5).

[12] For more information, see Drost in Martin Grambow (Publisher) "Nachhaltige Wasserbewirtschaftung" [*Sustainable Water Management*] under No. 4.2.1 Sustainable Legislation, SpringerVieweg Verlag 2013.

[13] The Treaty of the European Union, as laid out in the Lisbon Treaty of 13 December 2007 (OJ no. C306 p. 1, OJ 2008 no. C 111 p. 56, OJ 2009 no. C 290 p. 1 OJ 2011 no. C 378 p. 3).

All competences not conferred upon the Union in the treaties remain with the Member States (see Art. 5 (2) EU Treaty). Under the principle of subsidiarity, the Union will act only in areas which do not fall within its exclusive competence if and in so far as the objectives of the proposed action cannot be sufficiently achieved by the Member States, either at a central level, a regional level or a local level, but rather, by reason of the scale or effects of the proposed action, can be better achieved at Union level (see Art. 5 (3) EU Treaty). The bodies of the Union furthermore apply the principle of subsidiarity according to the Protocol on the Application of the Principles of Subsidiarity and Proportionality. National parliaments ensure compliance with the principle of subsidiarity in line with the procedure laid out in that Protocol (see Art. 5 (4) EU Treaty).

According to Art. 4 of the Treaty on the Functioning of the European Union (TFEU),[14] the principal area of environment is among the shared competences. According to Art. 191 (1) TFEU, Union policy on the environment shall contribute to the pursuit of the objectives: preserving, protecting and improving the quality of the environment, protecting human health and prudent and rational utilization of natural resources. Art. 11 TFEU stipulates that environmental protection requirements must be integrated into the definition and implementation of the Union's policies and activities, in particular with a view to promoting sustainable development.

The fundamental provisions of the European Union therefore stipulate the principle of subsidiarity as a guideline for all of the Union's action in environmental protection, and therefore also in water management.

2.2.3.3 National Provisions

Pursuant to Art. 20 (1) of German Basic Law[15] (GG), the Federal Republic of Germany is a democratic and social federal state. Art. 30 GG states that the exercise of state authorities and fulfillment of state tasks are vested in the federal states (German: Laender) unless otherwise stipulated by the Basic Law. In water management, the latter applies for federal waterways (see Art. 89 GG). According to Art. 20a GG, the State, also in conjunction with responsibility for future generations, protects the natural foundations for human and animal life within the scope of constitutional order via legislation and in line with the law via executory authority and jurisprudence. The regulation of the management of water resources is subject to concurrent legislation pursuant to Art. 74 (1) No. 32 GG. When it comes to concurrent legislation, Art. 72 (1) GG gives Germany's states the authority to pass legislation as long as and to the extent that the federal government has not made use

[14] In the version of the announcement from 9 May 2008 (OJ no. C 115 p. 47), amended multiple times.

[15] Basic Law of the Federal Republic of Germany ("German Constitution") in the adjusted version published in the Federal Law Gazette Part III, Outline Number 100-1, last amended via Article 1 of the Law from 23 December 2014 (Federal Gazette I pg. 2438).

of its legislative competence by passing law. Even if the federal government has made use of its legislative competence, Art. 72 (3) GG stipulates that Germany's states can adopt deviating regulations regarding the management of water resources (except for regulations related to materials or facilities).

National regulations on water management therefore primarily place the tasks with the states implementing the principle of subsidiarity. The federal government has discretionary authority for provisions superordinate to the federal states. However, this principle of "federal law over state law" pursuant to Art. 31 GG is correlated in favor of the principle of subsidiary within the scope of deviating legislation pursuant to Art. 72 (3) GG, with state law overriding federal law. The right to pass deviating legislation is within a state's discretion, such that the smaller unit ultimately decides which regulatory objects will remain on the superordinate level. Exempted from this, according to the will of the body that passed the Basic Law, are regulations related to materials or facilities, for which it is necessary to have regulation that is uniform on the federal level.

2.2.3.4 Regional Provisions

As the example of Bavaria shows, regulations in water management were largely laid out in state legislation for historical reasons. Water management regulations were codified for the first time in the Kingdom of Bavaria with the three water laws of 1852, the Law on the Use of Water, the Law on Irrigation and Drainage Activities for the Purpose of Soil Culture, and the Law on the Protection of River Banks and Against Floods. The Water Act for the Kingdom of Bavaria of 1907 carried on this legal tradition. State legislative regulations were also adopted for the realms of civil law, and therefore property law in regard to water [16]— deviating from the Civil Code regulations applicable across the Empire as of 1900. There was no regulation on water resources across the Empire. Only under the provisions of the Basic Law, with the Federal Water Act of 1957, entering into force on 1 March 1960, federal framework regulations were passed for the first time.[17] However, they were structured as guidelines for the federal states, otherwise leaving the detailed fulfillment of the provisions of water law up to the states.

Within the German states, the tasks of water management are distributed among the individual levels of state administration according to the respective organizational norms,[18] or individual tasks in water management are assigned to the local authorities, cities, market towns and municipalities. Accordingly Art. 83 (1) of the

[16] See Art. 65 of the Introductory Law of the German Civil Code.

[17] See Art. 75 GG, in the version applicable until 1 September 2006.

[18] See, for example, Bayern die Verordnung über die Errichtung und Organisation der staatlichen Behörden für die Wasserwirtschaft [*Ordinance on the Set-up and Organization of State Authorities for Water Management*] – OrgWasV from 4.12.2005 (Law and Ordinance Gazette p. 623), last amended by the Ordinance from 1.3.2012 (Law and Ordinance Gazette p. 86).

Bavarian Constitution transfers the task of water supply to the municipalities, the same applying under Art. 34 of the Bavarian Water Act (BayWG) to the task of sewage disposal, and under Art. 22 BayWG to the task of maintenance for smaller category III water bodies.

The regulations among Germany's states regarding water management therefore carry forth the federal legislative approach of implementing the subsidiarity principle, particularly as regards cities, market towns and municipalities in fulfilling their water management duties. Water management duties are selected from the overall task portfolio according to the capacity assumed of municipalities. The selection reflects proximity to the local community and therefore to citizen fulfillment of community duties.

2.2.3.5 Local Provisions

Local provisions in water management have to be made in order to fulfill tasks assigned by higher hierarchies. In the case of Bavaria, this especially applies for the tasks of water supply and sewage disposal, which municipalities must regulate in line with the requirements laid out in German legislation regarding municipalities.[19] A municipality's citizenry is obliged to use these facilities as well as to carry the burden of use and facilities.[20] Local law maps water supply and sewage disposal as community tasks, with citizens being obligated to contribute to the fulfillment of this task, yet also having to accept the standards set out by the local community.

Local rules on water management are mainly based on the relationship the local community has with the citizenry, governing the rights and obligations of that relationship. They also encompass financing of community facilities in order to fulfill the tasks via community charges, their collection being structured according to the principles of cost coverage and equivalence.[21] In relation to the state, municipalities are subject to a general governmental surveillance system that also has to be applied to the fulfillment of water management tasks.[22]

2.2.3.6 Individual Rights

Individual competences arise primarily from civil law, particularly from property law. The Federal Republic of Germany constitutionally guarantees property rights in Art. 14 (1) line 1 GG. However, this guarantee is contrasted by the social

[19] See e.g. Art. 24 of the Bavarian Municipal Code (BayGO).

[20] See e.g. Art. 21 BayGO.

[21] See, e.g. Art. 5 and 8 of the Bavarian Municipal Revenue Act in the version published on 4 April 1993, amended multiple times.

[22] See e.g. Art. 110 to 113 BayGO.

obligation of property pursuant to Art. 14 (1) line 2 GG. In regard to water management, individual rights are only specified to a limited degree.

According to the rulings of the German Federal Constitutional Court,[23] water management has overriding significance for the general public interest. This significance supersedes the right of property, with the consequence that the latter can only be asserted in a highly restricted manner. German law does not recognize water property for flowing bodies of water, nor for groundwater.[24] With few exceptions (proprietary use, common use), the interventions in water bodies as regards water use or development are subject to ("central") governmental approval.[25]

Although the right of access to safe drinking water and sanitation is designated as a human right and thus as an individual right of human beings, individual rules on water management only assume relevance when these tasks de facto cannot be resolved by the local community, or cannot be resolved due to disproportionality. The principle of subsidiarity in water management on the individual level is therefore characterized by a shifting of tasks to the next-higher organizational level. This process leads to a governmental guarantor's obligation making the state the third immune system of each individual.

2.2.4 Discussion Regarding the Forms and Further Development of the Principle of Subsidiarity in the Legal Requirements for Water Management

Taking into account the principle of subsidiarity, the general legal requirements for water management have found their way into legislation at the supranational, national and regional levels. Yet this has led to profound changes in the original hierarchical structure of subsidiarity. Thus, particularly at the supranational level, new approaches have been undertaken, separating the scope, extent and objective of water management from national and federal state related structures. More and more new organizational forms (joint bodies) emerged, which in turn needed to be integrated into existing governmental structures.

Water management under European Law is shaped by the Water Framework Directive (WFD).[26] Good ecological status in surface water, good ecological potential in heavily modified surface water, and good chemical status in surface water, good chemical status of groundwater and good quantitative status have been stipulated as environmental objectives (see Art. 4 WFD).

[23] Ruling of the German Constitutional Court of 15 July 1981, case no.: 1 BvL 77/78 – known as the Wet Extraction Ruling.

[24] See § 4 (2) WHG.

[25] See § 4 (3) WHG.

[26] Directive 200/60/EC of the European Parliament and of the Council of 23 October 2000 establishing a framework for Community action in the field of water policy (OJ. L 327 of 22 December 2000 p. 1), amended multiple times.

These environmental objectives should be attained in river basin districts (see Art. 2 (15) WFD). A river basin district encompasses a river basin from which the entire surface water flow meets at one single river estuary or sea delta via streams, rivers and potentially lakes. River basin districts can be divided into sub-basins, i.e. the land from which all surface run-off flows through a series of streams, rivers and, possibly, lakes to a particular point in a water course (normally a lake or a river confluence) (see Art. 2 (14) WFD). Groundwater is divided into bodies of groundwater as demarcated groundwater volumes within one or multiple aquifers (see Art. 2 (12) WFD), allocated to the river basin districts.

Thus, the guidelines encompass a new geometric categorization of the objective of water management, and therefore exist at least parallel to the current government structures in most Member States of the European Union. The potential subdivision of river basin districts into sub-basins or in bodies of water should not be used to adjust the management object of "bodies of water" to match government administrative structures, rather merely delimit the scope of the management object and render the requisite measures for improving, hindering deterioration or monitoring compliance with the requisite measures for attaining the environment objectives manageable and transparent.

Management plans and measure programs established on the river basin level yield, depending on the set-up process, secondary supra-local, supra-regional or even supranational structures that do not correspond to the primary political subsidiary pattern. Thus, meta-structures are created that partially exist parallel to the political hierarchical structures, and additionally are organized "transversely". For example, "management" of a supranational river basin district requires independent management structures laid out in agreements between the states involved (see e.g. Elbe River Basin Community; German: Flussgebietsgemeinschaft Elbe). These water body sub-structures stipulated by the Water Framework Directive do not have a direct hierarchical relationship among each one another. Rather, they simply complement the aggregate system of a river basin district. Thus, as regards the management object, water management does not initially follow the principle of subsidiarity. It corresponds to a subsidiary application of the supranational approach, in which sovereign states bind themselves to a meta-structure.

As regards water management, Art. 3 WFD stipulates that Member States define their sections of river basin districts and ensure via management agreements, including the definition of appropriate competent authorities, that the WFD is implemented in every river basin district. This promotes water management administration on an equal level between the Member States, an approach corresponding to the relationships of the Member States of the Union among each other.

However, the Water Framework Directive does not regulate the organization of the water management within the Member States. In this regard, the principle of subsidiarity applies under Art. 5 (3) of the EU Treaty. Their realization remains up to national regulations.

No uniform federal administration has been created for the river basin districts crossing state borders. A corresponding amendment to the Basic Law would have been necessary to that end pursuant to Art. 30 GG. Rather, the administration of water management is fit into the governmental structure of the Federal Republic of

Germany, which is influenced by the principle of subsidiarity. Coordination of water management among Germany's states as well as with other Member States of the European Union, that share territory in a river basin district, should occur on the level of administrative agreeagreements.[27] Thus, water management towards other Member States of the European Union is not considered to be international affairs, for which Germany's federal government would have exclusive jurisdiction (see Art. 73 GG).

Although restructuring water management into river basin districts, sub-basins and bodies of water is implemented into national law, the formation of intra-state administration of water management is still subject to regulation based on subsidiarity between Germany's federal government and states, which, however, is supplemented with interactive, potentially supra-sovereign components.[28] National and international "joint bodies" have now been created in all river districts, harmonizing management and producing mutual reports and measure programs for the river basins.

2.2.5 *The Principle of Internality*

As shown, subsidiarity is the underlying principle for the collaboration among the Member States of the European Union. In the important sector of water management its application is however limited, as the Member States are obligated to implement the Water Framework Directive into national legislation and Art. 5 (3) of the EU Treaty allows Member States to deviate from the principle of subsidiarity. The requirement of water management based on river basins is system neutral when it comes to designing the administration of water management.

The Water Framework Directive and its approach based on natural conditions can nevertheless serve as an example for setting up decentralized structures deviating from the principle of subsidiarity in order to fulfill tasks. The structural approach of the Water Framework Directive should be contemplated with respect to internality, with its principles of preventing further deterioration and the need for improving and/or maintaining good ecological and chemical status of water bodies and/or at least the good ecological potential for bodies of water where the restoration of good ecological status would be disproportionate to the level of river basin districts.

Internality is the conviction that the results of action tend to be experienced as being self-caused and self-induced instead of being attributable to external causes.[29] The term originates from psychology and of course cannot be directly applied to objects like bodies of water. In a figurative sense, however, the

[27] See § 7 (3) (1) WHG.

[28] See the remarks on the "joint bodies," in Ute Mager 2010 (Uruguay v. Argentina).

[29] Schuler et al. (2001).

requirement of "good natural status" as a reference status and therefore an objective to be pursued pursuant to the WFD is the dedication towards self-conduct and input of resources. Thus, deviations from this status are based on self-responsibility. When it comes to water management based on river basins, this translates to self-responsibility among the main parties responsible for them. The conviction that deviations from good status as an environmental objective are the result and consequence of their own conduct and therefore need to be corrected under their own responsibility will ultimately lead to structures that can be used for objective attainment.

Irrespective of the sector of water management, it should be noted that tasks within the scope of decentralized subsidiary or miscellaneous centralized non-hierarchical structures can only be set to the extent a decentralized unit is capable of fulfilling those tasks on its own merit with internality. Conversely, decentralized structures can be created based on the conviction of the parties involved that they are capable of fulfilling a task internally with self-responsibility.

2.2.6 Subsidiarity and Internality Within the Scope of Future Development of Decentralized Structures

Subsidiarity can ultimately be understood as a hierarchical top-down process in which central governmental order and accountability are delegated back to a lower hierarchical level ("back" because legitimacy itself ultimately always originates from the people, and therefore comes from the individual and local structures). The reverse process, i.e. bottom-up, of passing sovereign responsibility to a meta structure, which means all "supra-" solutions, either leads to a new central starting point for an additional hierarchical order (see the development of the Federal Republic of Germany out of the small states of the Post-Napoleonic Period) or to a parallel or meta structure of sub-sectors or certain processes.

While in the latter case transparency and the respective parliamentary legitimacy can indeed become a challenge, this kind of structures tend to be on the rise in Europe within the scope of decentralized approaches. For example, river basins are not only gaining importance as a decentralized approach in water policy—which is physically justifiable in this case—but also serve as the beginnings of supra-sector cooperation projects, such as the Danube Strategy, or as a sea basin in the Mediterranean Strategy and Baltic Sea Strategy. As the youngest sprout in the meta-structures, the Alpine Region Strategy transgresses both river basins and sea basins. One finds a presentiment of similar developments for the river basins in the papers published by the UN and its sub-organizations.

In this context, subsidiarity is a political principle that, as shown by the example of the European Union, is primarily derived from pre-existing structures (in this case, the Member States). Meta structures formed outside of the principle of subsidiarity are generally based on the ordering principle of internality. They map out the framework under which tasks can be performed with self-responsibility and autonomy in line with a specific objective. The principles of subsidiarity and internality are not mutually exclusive. Rather, they complement each other. Under the principle of subsidiarity, internality has to be taken into account and implemented on every level of task fulfillment. Shifting responsibility "up" or "down" and the resulting allocation of problems to external causes does not ultimately lead to objective attainment on the subsidiary level. This will also apply intersectorally and intertemporally.

However, the example of water management in subsidiary and miscellaneous meta-structures also shows the necessity of structuring them in a dynamic manner. In particular, the objectives of the use of water have to be adjusted in an ongoing manner to match the changes that take place in a society's needs. Nevertheless, this approach of internality provides a certain degree of stability, as it only permits deviations within a certain bandwidth of self-responsibility through objective definition—within the scope of water management, one example here is the reference to "good status". In the system of subsidiarity as well as in decentralized non-hierarchical systems, self-responsibility has to be carried and practiced on all levels and by all parties involved. This requires a high level of communication and participation. Once that is ensured, structures based on the principles of subsidiarity and internality can be created for fulfilling tasks and solving problems. They can also be dismantled once they are no longer necessary.

It can be assumed that this diversity in structures is one of the potential answers to the post-national statehood induced by globalization.

2.3 Smart Energy Systems: Decentralization, Virtualization and Hierarchization

Bernhard Schätz

2.3.1 Key Messages

The need for the increased integration of renewable resources on the one hand, and for the reliable availability of electric energy on the other hand are driving forces for the construction of an "intelligent" grid, ensuring the necessary resilience concerning imbalances of demand and supply.

2.3.2 Renewables and Reliability: A Contradiction?

The reliable supply of electric energy has become a commodity, on which developed countries rely, turning it into one of the backbones of modern society. In Germany, for example, the average outage per user and year amounts to 16.92 min (Bundesnetzagentur 2014). Unreliability in energy supply is—especially in high-technology countries—a serious economic issue. By relating the use of electric energy to the creation of economic value and thus defining the "value of lost load" (VoLL) For example, as shown in Fig. 2.9, an outage of 1 kWh corresponds to an economic loss of at least 5 € in the majority of Germany (Piaszeck et al. 2013).

However, with the quest for a sustainable energy production, the need for the use of renewable resources increases. Figure 2.10 shows the example of Germany, with an increase in percentage from 6 % to 23.5 % of renewables with respect to the gross energy demand in little more than a decade (BMU 2013). Furthermore, the highly volatile sources—especially wind power and photovoltaics—constitute more than half of the renewables. Furthermore, these sources are among the fastest growing–indicating a further increase in volatility in energy production in the future energy grid.

As a consequence, the electric grid is faced with two contradicting goals: On the demand side, a high flexibility is expected with respect to providing sufficient

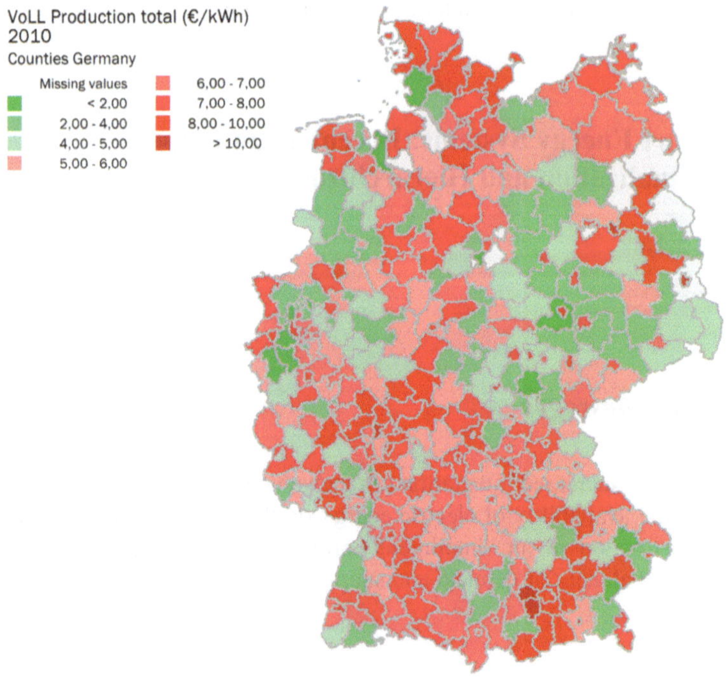

Fig. 2.9 Economic loss through power outages (Piaszeck et al. 2013)

	Hydro-power	Wind-power Land-based	Wind-power Off-shore	Bio-mass	Photo-volta-ics	Geo-ther-mal Power	Sum Production	Percentage of Gross Consump-tion
	[GWh]						[GWh]	[%]
2001	22.733	10.509	0	5.207	76	0	**38.525**	**6**
2002	23.124	15.786	0	6.038	162	0	**45.110**	**7.7**
2003	17.722	18.713	0	8.841	313	0	**45.589**	**7.6**
2004	20.095	25.509	0	10.471	557	0.2	**56.632**	**9.3**
2005	19.638	27.229	0	14.354	1.282	0.2	**62.503**	**10.2**
2006	20.008	30.710	0	18.700	2.220	0.4	**71.638**	**11.6**
2007	21.170	39.713	0	24.363	3.075	0.4	**88.321**	**14.2**
2008	20.443	40.574	0	27.792	4.420	17.6	**93.247**	**15.1**
2009	19.031	38.610	38	30.578	6.583	18.8	**94.858**	**16.3**
2010	20.953	37.61	174	34.307	11.729	27.7	**104.810**	**17.0**
2011	17.671	48.315	588	27.603	19.599	18.8	**123.775**	**20.4**
2012	21.793	49.948	722	43.550	26.350	25.4	**142.418**	**23.5**

Fig. 2.10 Increase of renewable resources in Germany

energy when requested, especially during peak times like the evening hours in case of households, as shown in Fig. 2.11. On the supply side, a high flexibility is expected with respect to accepting sufficient energy when produced, especially during peak times like the noon hours in case of photovoltaic energy.

The electric grid itself basically can only provide means of transmission and distribution. It therefore has not the capability to store energy, and act as a buffer between times of high demands and low supply on the one hand, and low demands and high supply on the other hands. As a consequence, in the traditional approach, a small up to medium number of high-volume facilities—like fossil or nuclear power plants—are used to balance supply and demand by running those few plants according to a demand-driven load profile. However, the increased use of renewable resources has lead to a proliferation of smaller, widely distributed and often privately owned facilities, like household photovoltaic installations or communal wind turbines. Obviously, the centralized coordination scheme (generally requiring a substantial amount of manual ahead-planning and human intervention)—perfectly suitable for a small number of high-volume facilities—unfortunately is no longer adequate for a large number of low-volume facilities.

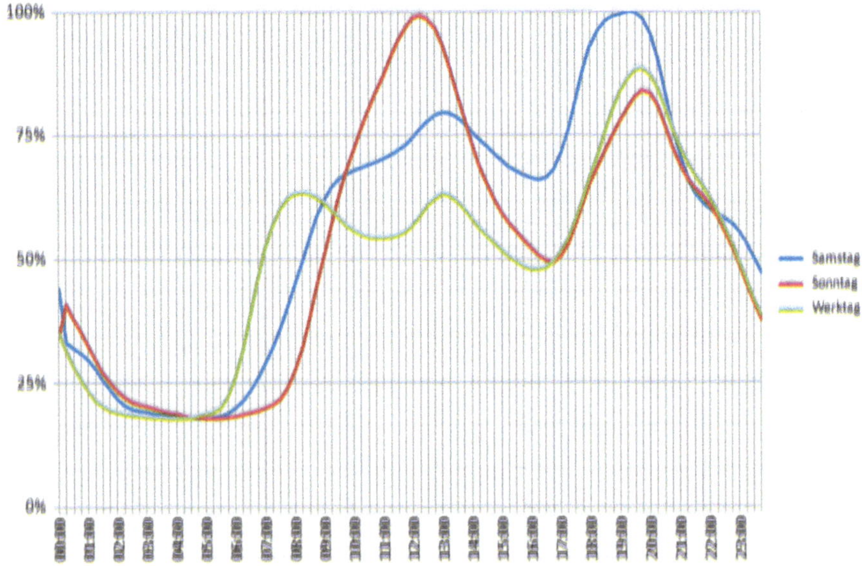

Fig. 2.11 Standard load profile for different weekdays of a German Household

2.3.3 *Smart Distribution: Virtualization, Decentralization, Hierarchization*

In its 2011 Position Statement ("Eckpunktepapier") "Smart Grid und Smart Market" (Bundesnetzagentur 2011), the German Bundesnetzagentur provided several theses with respect to the aspects of a changing supply system for energy. Among others, the decentralization is seen as a driving force of change. This is—at least in Germany—cause by the changing use of the grid. The classical distribution grid no longer only distributes medium and low voltage energy provided in upper voltage layers to the end user, but rather takes over the new task of accepting and distributing energy volumes produced in decentralized installations, and also exporting them to the upper voltage layers. To achieve a manageable and stable control of such a system, a "cellular" approach is suggested, using self-controlled interacting structures. Essential element of such an approach is the implementation of "self-regulating structures (cells) which are parallely or hierarchically aligned" where each cell acts independently, achieving a "regionally or locally optimal balance of demand and supply".

The use of hierarchical structures supporting a decentralized control by means of a "fractal" (i.e., self-similar) cooperation scheme is a well-established paradigm in the ICT domain. This kind of architecture revolutionized, for instance, the concept of (electronic) communication, forming the foundation of the Internet, which is therefore also called "network of networks". Building upon three central principles—*virtualization, decentralization*, and *hierarchization*, which in turn

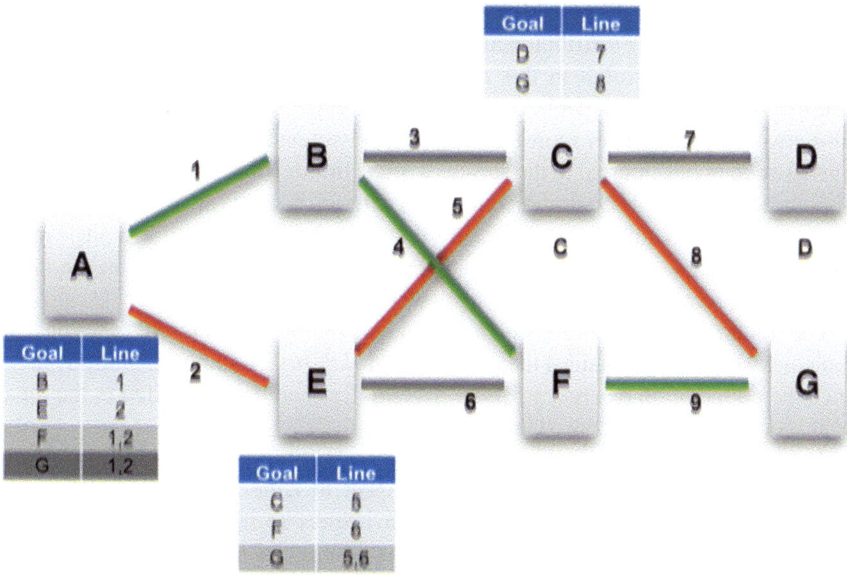

Fig. 2.12 Virtualization and decentralization in an Internet network

will be shortly illustrated for the case of the Internet protocol family (Tanenbaum and Wetherall 2013)—allows to substantially increase efficiency and robustness of large-scale complex systems like the communication system.

In the Internet context, *virtualization* is achieved by breaking up a dedicated point-to-point line between the sender and receiver into a set of paths of linked neighboring nodes starting with the sender and terminating with the receiver, as well as breaking up a (prolonged) communication into a set of individual packages. The virtualization, which is achieved by these means of *fragmentation*, increases *flexibility* since a (virtual) communication line can be implemented by simultaneously using alternative paths over linked neighbors, and thus local failures or local congestions of the network can be immediately compensated, ensuring a continued undisturbed global service provision. Figure 2.12 shows an example of such a fragmented topology, illustrating two alternative paths (links 1–4–9, and 2–4–8) between nodes A and G.

Decentralization in the communication network is achieved by substituting a dedicated node responsible for the definition of possible communication paths between a sender a receiver by a distributed approach, where each node iteratively collects from and—after aggregation—in turn propagates to its neighbors information about which node is (directly or indirectly) reachable along which link, potentially with what quality. The decentralization, which is achieved by the *cooperation* between all nodes of a system, enables the *stability* of the system, since the nodes autonomously adapt their routing information when topology changes like addition or removal of nodes or links occur. Figure 2.12 shows such an example of a network with partial routing information for nodes C, E, and A at different stages of this distributed algorithm.

Finally, *hierarchization* in this context is achieved by structuring larger networks into networks of networks by means of aggregating parts of the network into sub-networks with (one or more) dedicated gateway nodes, such that communication between nodes internal to the sub-network is limited to paths completely within this sub-network, while communication with external nodes is always routed via the gateway nodes; as this principle can furthermore be repeated on a high level—turning a complete sub-network of a lower level to an abstract node on the next-higher level, a fractal hierarchy is established.

The fractal *hierarchization*, which is achieved by the delegation of functionality to the next-high level, enables the global *controllability* of all nodes in the system, since the cooperation necessary for the decentralized control is limited to the most local scope, avoiding inconsistent decisions.

An architecture providing these capabilities—flexibility, stability, and controllability—offers substantial benefits with respect to *robustness of operation and management:* It *autonomously adapts*—with respect to the former—to transient disturbances like local congestions or failures of links or nodes, as well as—with respect to the later—to the static addition and removal of nodes or links. The application of these architectures in modern information and communication technologies have demonstrated the ability to substantially increase robustness and adaptability of systems, as seen in large-scale robust internet-based functionalities like Voice-over-IP Telephony.

By using the so-called "embedded systems" technology, i.e. digital information processing combined with analog electric/electronic control, as a bridging technology, these information and communication architectures are increasingly incorporated into large-scale systems governing physical (i.e., mechanical, chemical, electrical, etc.) processes on the one hand and "cyber" (i.e., organizational, economical, etc) processes on the other hand. By exploiting the above-mentioned abilities not only for the communication and information infrastructure, but using them to implement governing schemes for these processes, these 'cyber-physical systems' (Geisberger and Broy 2012) can improve the robustness of traffic systems, production facilities, or electric grids.

By applying the principles of virtualization, decentralization, and hierarchization of the communication grid in an analogue fashion to the electric grid, this paradigm of hierarchical cooperative structures—built around the *concept of low-volume energy "packets"* balanced and aggregated in a cellular fashion before begin exported to upper smart grid layers—offers the mentioned advantages for energy systems. By turning the traditional distribution scheme of electrical grids into a "internet of energy", and thus implementing the cellular approach mentioned above, a similar robustness can be achieved, which is—as argued above—specifically necessary given the increased integration of renewable resources.

2.3.4 Smart Grid: Resilient Balance of Demand and Supply

As shown in Fig. 2.13, to implement such a smart distribution grid, the electric grid is over-laid with a communication network by enhancing each grid node—for example, a household or commercial building, potentially also equipped with a photovoltaic installation or battery buffer—with communication as well as sensor—for example, intelligent meters—and actor—for example, inverters—device technology. Furthermore, control software is used to execute decision rules controlling these actor devices based on information provided by the sensor devices, improved by algorithmic predictions of the energy demand or supply of this grid node. Additionally, the predicted local supply and demand can be exchanged with the grid via communication devices, receiving global data in return, to optimize the load control.

By these means, each node is turned into a cooperating agent, negotiating its energy demand and supply within the grid. To achieve the fractal hierarchization, the network is divided into sub-networks, where each sub-network is assigned a dedicated coordinating agent, managing the negotiation within the sub-network—aggregating the demands and supplies as well as balancing them as much as possible by assigning individual plans to the nodes—and handling the negotiation with the rest of the network—by communicating the aggregated demand/supply information and in turn accepting a fitting plan for the sub-network. To provide the necessary information for a negotiation, each node provides a plan describing the expected demand requested, and supply or buffering capacity provided for a defined period—for example, 24 h—and fragmented into small slots—for example, 15 min—to the coordinating agents. To implement a balanced plan, each node

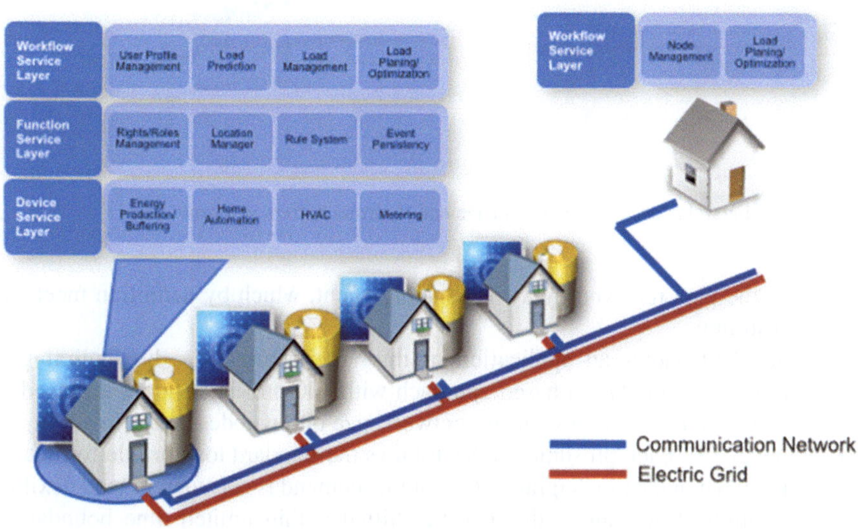

Fig. 2.13 Architecture of a decentralized smart energy system

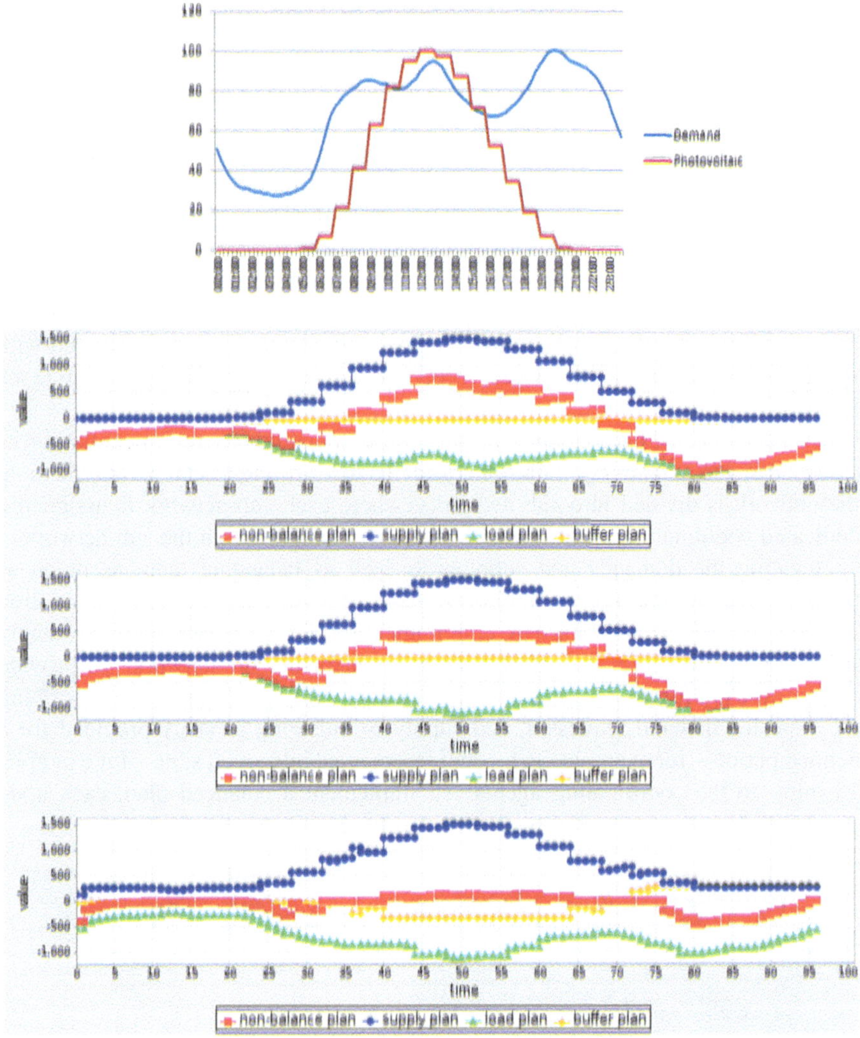

Fig. 2.14 Hierarchic decentralized management of virtual buffers in a smart grid

executes the plan received by a coordinating agent, which by definition meets its own constraints.

Figure 2.14 shows an application of this architecture to a distribution grid scenario using a simulation-based approach with 100 prosumer nodes arranged in a three-layered architecture with sub-network size of ten nodes on the lower layers (Kießling 2013). Each prosumer nodes follows the standard load profile for households as shown in the top diagram. 10 % of this demand is considered to be flexible, like operation of appliances that can be shifted within limited time boundaries. Furthermore, 25 % of the prosumer nodes are equipped with a photovoltaic installation, with a supply profile of a factor 5 to the profile as shown in the same diagram. Additionally, 10 % of the prosumer nodes are equipped with a buffer storage capacity corresponding to 20 % of the supply, while another 10 % is equipped with small-volume heat-power co-generators. The three lower diagrams show the results from different balancing strategies executed by this fractal architecture. The first diagram shows the balancing without flexible loads or buffers, the second diagram including flexible loads, and the third diagram including the use of the buffers and heat-electricity co-generators. Using such an fully autonomous approach, the system is capable of drastically reducing the load imbalances—which are indicated by the red line—in the first diagram, which generally require the use of additional (conventional) facilities, leading to a nearly balanced load in the third diagram.

To realize the necessary cooperation, as shown in Fig. 2.13 each grid node is equipped with the capability to predict, optimize, and manage the production and consumption load of this node. To that end, it makes use of a (modifiable) set of rules defining the actions—like turning on or off electric devices or switching between charging and discharging the battery buffer—to be performed based on internal—like user requests or status changes of devices—or external events—like load shift requests (Koss et al. 2012). Nodes only acting as an aggregator for sub-networks may only use a subset of these functionalities to support the hierarchization of the distribution gird.

2.3.5 Conclusion

While, as stated in Faulstich (2012), the increased introduction of renewable resources will lead to a degradation of the power supply unless "each factory, office, or store measures and controls the power flow at each instance and power predictions will become daily routine", the major challenge lies not in each of those tasks individually, but rather in their global coordination. Here, the approach of cyber-physical systems offers a suitable solution. However, these system will differ substantially from current systems, exhibiting the properties "cross-X"—for example like cross-organization, cross-domain, cross-discipline—as well as "live-X"—for example, life-update, life-reconfiguration—and finally "self-X"—for example, self-monitoring, self-healing—not found in traditional systems (Schätz 2014). Using these capabilities cyber-physical systems supports the construction of architectures coordinating large-scale processes robust against and adaptive to unexpected or changing behavior of their users and environments, based on the principles of virtualization, decentralization, and hierarchization. The practicably of this approach for the electric grid is currently demonstrated in first solutions for energy distribution, where the above-mentioned principles allow the stable integration of low-volume renewable resource like wind energy or photovoltaics using the hierarchic aggregation of those low-volume productions—both in an industrial setting like in "virtual power plants", or in more experimental settings like connected living-labs using a market-based approach to decentralized buffering and load-shifting schemes including an incentivation for cooperative behavior. By implementing the 'cellular' approach for grids and markets asked for by the German Bundesnetzagentur in its 2011 Position Statement 'Smart Grid und Smart Market', these cyber-physical systems architectures allow to solve the conflict between sustainability and stability of the provision of electric energy.

2.4 The Importance of Negative Feedback

Martin Korndörfer

2.4.1 Key Messages

The author postulates that many if not all of the problems faced by the modern global society can be traced back to the all-pervading centralised approach of human activities and the subsequent deterioration of adequate feedback loops between natural and anthropogenic as well as within anthropogenic systems.

2.4.2 Introduction

Natural systems follow Holling's model of an adaptive cycle (Fig. 2.15). Disturbance of a climax system K leads to a release (Ω-phase) of resources and makes them available for pioneering species that make use of those resources for reorganisation (α-phase) and exploitation (r-phase). As ecological niches are filled, more and more specialised species develop filling niches that through this specialisation become available 'between' existing niches ($r \rightarrow K$). This is the conservation phase of the adaptive cycle, changes happen slowly and the system as a whole is stable until a perturbation re-starts the cycle.

In natural ecosystems no intentionality, no foresight and no values exist and the influence of each agent on the whole system is relatively low. Perturbations happen in regular (e.g. seasonal flooding) or stochastical intervals (e.g. earthquakes, volcanic eruptions).

In anthropogenic systems the same cycles exist. In some cases these cycles follow natural cycles of disturbance (anthropogenic systems rely on natural systems), where drought, flooding or some other natural disaster cause anthropogenic systems to fail. In other cases the anthropogenic system itself becomes maladapted and has to change. In anthropogenic systems this change is often triggered by technological innovations (e.g. modern telecommunications) and/or public opinion

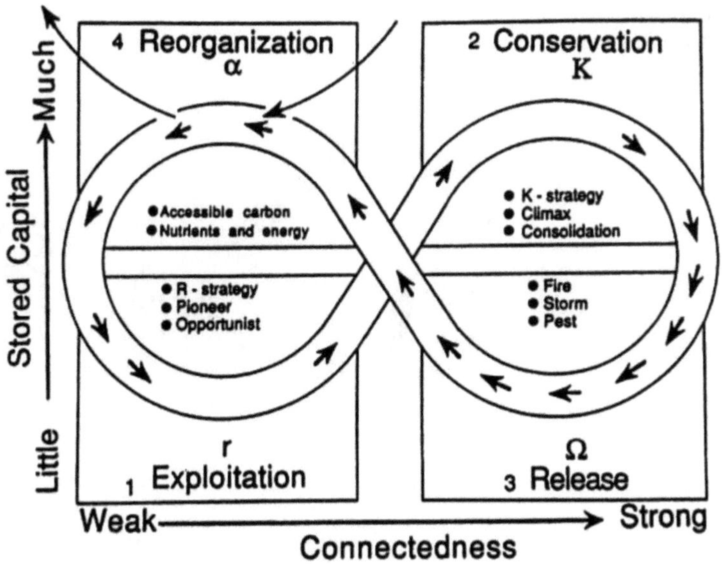

Fig. 2.15 The adaptive cycle according to Holling (1995)

(e.g. educational system, adherence to human rights). Anthropogenic systems are influenced by foresight, vision, values, psychology and other factors unique to the human species. In addition, humans have invented technologies that greatly multiply their individual and collective impact on anthropogenic systems (e.g. via television or advertisement) and the earth system (e.g. acid rain, climate change, mining, fracking). As humans are increasingly able, through technology and understanding of the earth system, to affect natural systems and natural cycles, the functioning of the earth system increasingly depends on the anthropogenic systems being sensitive and adaptable to natural feedback loops (e.g. controlled burning to avoid wildfires in Australia).

2.4.3 Capital and the Neoliberal Monetary System in the Context of the Adaptive Cycle

The role of capital: We cannot hope to find solutions to the challenges posed by 'too much' centralisation if we do not acknowledge the centralisation-favouring effect of capital and our current economic system. In the not too distant past wealth and power was determined by the amount of land that was controlled by a person or a people. There was a natural limit to how much control and power a single agent (person, government, business) could accrue. While the monetary system still had its basis in real-world assets there was also a natural limit of how much wealth and hence power a single agent could accrue: the spatially distributed amount of

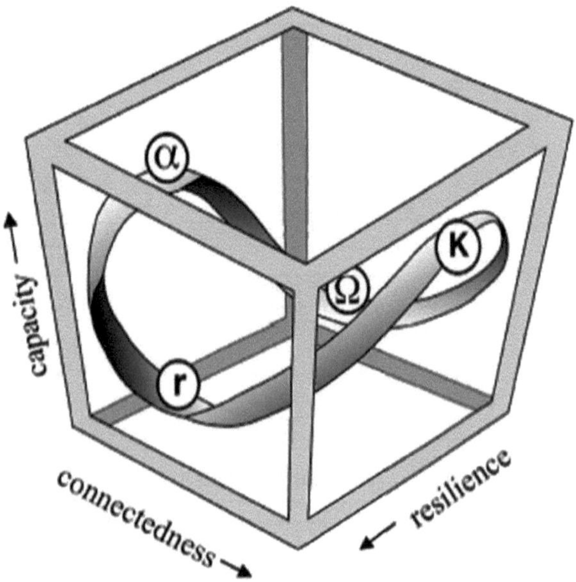

Fig. 2.16 Three dimensional model of the adaptive cycle (Holling et al. 2002)

physical resources available and controlled. Now, that the creation of money has been largely decoupled from real-world assets, a natural limit to the level of centralisation of wealth, power and control has ceased to exist. This allows the individuals and societies in power to keep the anthropogenic global system within the K-phase of the adaptive cycle even as the system rapidly loses resilience and becomes ever more brittle (Fig. 2.16).

Western societies therefore risk being trapped in a 'rigidity trap', while those exploited societies and ecosystems in other parts of the world often end up in a 'poverty trap' (Fig. 2.17).

While capitalism favours centralisation of power at the same time it exerts this power in a decentralised manner via the monetary system. This way a large number of independent actors do not need to be controlled individually but follow more or less on their own accord the rules set for the system as a whole. The overarching theme in capitalism is that each individual strives for the accumulation of capital, which can be re-invested in an economy that relies on growth to provide returns to those investors. The broad aim of the market economy to provide goods to people in the most efficient matter by employing free markets and regulation of demand and supply through prices has been narrowed down to a focus on profits and a growing economy (Felber 2010; Paech 2012). It is mostly this increasing single-mindedness of the capitalist system with an increasingly centralised control, which is responsible for the anthropogenic system becoming more and more insensitive to natural feedback loops (e.g. climate change, biodiversity loss, desertification, overfishing etc.) but also to feedback loops within the anthropogenic system itself (e.g. rising numbers of refugees, rising inequality, popularity of terrorist organisations, ineffectual demonstrations etc.).

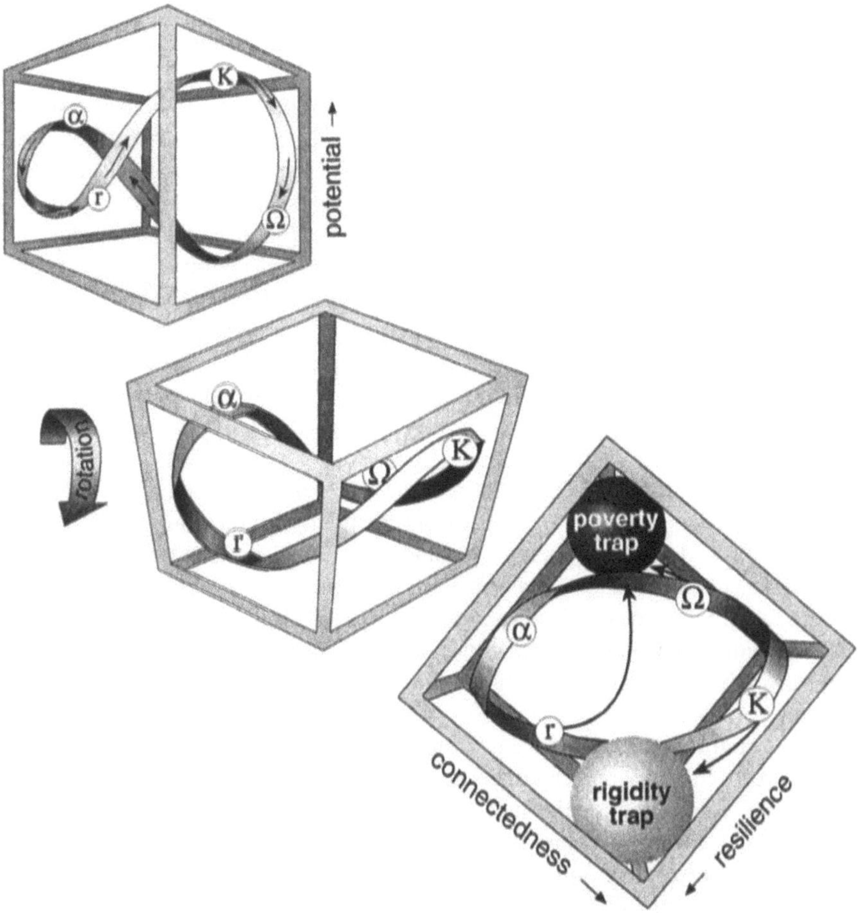

Fig. 2.17 Maladaptive systems trap anthropogenic or natural systems in alternative stable states, where systems are stopped from undergoing the necessary steps of the adaptive cycle (Holling 2001)

One example where the anthropogenic system has been caught in such a rigidity trap is Germany's now rapidly changing energy production system. Energy production systems of the past were characterized by centralisation—no great diversity of technologies or fuel and centralised ownership/control. A decade ago, there were only four major companies providing electricity to German households and industry. Those four companies basically split up the electricity market between themselves. Even with an enormous anti-nuclear community among the German population and proof that global climate change was man-made, and dropping prices for wind farms and solar power, these companies kept to their key business—nuclear power and coal. Instead of using their enormous capital, knowledge and man-power to spearhead the development of renewable energy systems, they lobbied heavily against a switch towards renewables and for the use of fossil fuels. Through their power and influence they tried to keep the system in the K-phase of the adaptive cycle, even though feedback from climate and population demanded adaptive change. In this instance these companies could not keep control over the system and lost a great part of their market share to decentralised renewable energy producers (Bontrup and Marquardt 2015). While in the centralised system of only four major energy providers most of the accrued capital was lost to the regions and cities through power fees, with a decentralised system in place a major part of the value added and purchasing power stays within the region or city (Breyer et al. 2014; Plankl 2013). In the new decentralised system, smaller organisations are leading the way, which are more sensitive to the needs and wishes of the population (often people from a region group together in democratically organised energy cooperatives). These smaller organisations are more agile in the market and are generally quick to adopt new, locally adapted business concepts (i.e. EnergieGenossenschaft-Flensburg (EGFl) was one of the first to trial the new power to gas technology). Decentralised power generation also has the benefit of needing less energy transport systems from the point of production to the point of use. Feedback loops of energy demand and local production capacity are re-established, which is why many regions trying to be self-sufficient in terms of energy are simultaneously also promoting energy saving measures. Finally, having many decentralised players means that the political influence of each individual company is limited (none is 'too big to fail') and are as such subject to democratic control.

This example shows that decentralising power/capital structures can contribute to an increase in democracy, an increase in public awareness, a fairer distribution of wealth etc. Again, the spatial and social context is paramount. Usually energy cooperatives do not have more than a few hundred members and are usually only active in their respective regions. On the one hand, the individual cluster of people and capital has to be sufficiently large and organised/cohesive to escape the poverty trap (see above) but not large or cohesive enough to be caught in the rigidity trap. For every system the optimal balance differs, in case of energy production systems with renewable energy the greatest benefit is realised with the greatest decentralisation, although there are even differences within these production systems. For example, more capital, people and organisation has to be accumulated before a wind turbine can be realised, while for photovoltaic systems much less of these resources are needed before the poverty trap is left behind and a project can be realised.

2.4.4 Anthropogenic Systems Emulating Natural Systems

Before presenting my conclusion on the main topic of centralised systems vs. decentralised systems, I want to come back to one of the central questions 'Does nature provide any practical guidelines'. In my opinion the answer has to be an emphatic 'Yes' as well as the side note that the processes by which natural systems are controlled follows the same basic rules as anthropogenic systems. The mostly artificial differentiation between the two is useful for discourse, but in the end it stems from our cultural heritage of understanding man as different from nature. Natural as well as human systems undergo cycles of disturbance-pioneering-innovation-adaptation-focussing-specialisation and, again disturbance. The resilient climax ecosystem that forms during and by this process has done so through emergence—the sum of all complex interactions of all agents involved (a decentralised process)—acting individually to actualise their own potential. Unlike natural systems, humans have given varying amounts of value to different parts of the system and aim to maintain the system in a state that favours their value systems. Hence, people with the means to avoid disturbance or save themselves from the effects of disturbance will do so, often at the expense of those lacking these means, especially when there is a significant lack of experiential/adequate feedback between the two groups or the system has fallen into a rigidity trap that is insensitive to feedback. The centralisation that we can observe in the current form of globalisation has not balanced the scales between those with the means to avoid disturbance and those who can't, but rather accentuated the unequal distribution of means (Stiglitz 2003). The last century has seen both a rise of control mechanisms to cope with this (insurances, access to lawyers, access to credit etc.) and their decline, mostly through privatisation (centralised systems) and the prerogative for growth and to increase profits. This has turned their initial intention (buffering individuals and the system from disturbance) upside down.

Size and complexity matter for the efficiency and control of centralised and decentralised systems. In the preparation of the 'Wildbad Kreuther' workshop it has often been proposed that centralised structures provide more efficiency while decentralised structures provide more resilience. I consider this assumption highly debatable. In nature, simple human societies and complex modern enterprises we can find centralised organisation only up to a certain size, above that, ants will form a separate colony, plants will outsource functions to symbiotic organisms, hunter-gatherer societies form a separate clan, corporations outsource parts of their supply chain to separate companies and their sales and after-sales services to yet other businesses. However unlike in natural systems, in anthropogenic systems (e.g. companies, political parties) there is a tendency of the 'mother' to retain a high level of power over the 'daughter'. These examples show that whether efficiency is a function of centralisation depends on the size of the system and its agents.

Whether anthropogenic systems have better efficiency than natural systems could also be argued, given the recent advances in biomimicry that show clearly that naturally evolved solutions outperform human engineering and design efficiencies (e.g. rotor blade for small wind turbines, chemical sensing abilities of some insects, spider silk) to a degree that we with all our technology struggle to simply replicate what nature already does, purely with renewable energy at ambient temperatures.

Also, the thesis that anthropogenic systems are superior in terms of control is, I believe, a misrepresentation. I believe that anthropogenic systems have the natural tendency to become more and more centralised. In order to maintain the centralised nature of the system and to control it, the system has to be reduced to include only a few controllable factors. Individual actors have to be controlled and brought in line (either directly or by proxy), in order to maintain system integrity. As the system is anthropogenic and follows the value system of one or a number of individuals in power, the system also follows only one particular value system—and due to the amount of power and control exerted can deliver the one output of that value system at high efficiency. This efficiency, however, comes at the cost of other co-outputs that could have been possible had a more complex system been employed instead of one with reduced complexity which is showcased in mono- and multi-cropping practices. If we agree that we need to embrace complexity in order to solve complex problems, we need to employ decentralised systems. For decentralised systems to work efficiently, system imminent feedback controls are necessary and would have to be fostered and protected by the new super-structure.

Recent research has shown that the capacity for meaningful feedback and self-regulation of human communities is limited by the human cerebral capacity for making and maintaining relationships with other human beings and lies at about 150 individuals. This could be an indication of how large each centrally organised cell of a larger decentralised system should ideally be, given that meaningful self-regulation, empathy, feedback mechanisms etc. are fundamental to our wellbeing and the functioning of our society (Hill and Dunbar 2003).

The last point for discussion that I want to put forward here is whether it even makes sense to pursue efficiency for its own sake, even if we accept that centralised systems are needed for better efficiencies. If we dare see the world as a place of abundance (abundant renewable energy—10,000× the worlds energy demand) abundant primary production, even abundant space, if we integrate human and natural habitats, would it not make more sense to pursue effective solutions rather than efficient solutions? All parts of a natural system perform multiple functions at the same time. Human systems are usually reduced in complexity and uni-functional (PV park OR pasture OR wildlife habitat) and hence use a lot of resources and occupy large amounts of space.

2.4.5 Conclusion

From these thoughts follows my personal conclusion about the proceedings of the workshop. In order to create a world that is sustainable, ecologically, economically and socially, we have to (re-)establish controlled (no boom and bust scenarios) ways to undergo continuous iterations of the adaptive cycle within the anthropogenic system. As the anthropogenic system exerts such control over natural systems but at the same time depends on intact natural systems, such a system would be -would have to be- sensitive to natural and societal feedback loops in order to avoid maladaptation. In fact, such feedback loops would have to be built into the system. Imperative for a successfully running anthropogenic system would be an awareness of the human psychology and limitations which determines the structure of the anthropogenic system varying degrees of central control in each cell (of adequate size) of a federally organised super structure. Mechanisms to control power must be established as the basic rule set of the system, where negative feedback loops are in place to avoid a runaway system towards serving increasingly a single agenda. Furthermore, the total dependence of the anthropogenic system on the natural system would have to be recognised and effective stewardship of the earth system embedded into the structure of the emerging sustainable anthropogenic system. In a similar way to the emerging decentralisation of Germany's energy production system, other malfunctioning anthropogenic systems, the financial system, the waste disposal system, the transport system for goods and people, the agricultural system etc. could and should be deliberately designed to become more decentralised, although the optimal level of decentralisation has to be established for each sector individually (what is the lowest limit of resource quantity (capital, level of organisation) to realise a given project and what is the minimal spatial extent that the project is realisable in).

References

References of Section 2.1

Arnold GW (2011) Challenges and opportunities in smart grid: a position article. In: IEEE proceedings, special issue: the electric energy system of the future, vol 99(6), June 2011

Diamond J (2005) Collapse: how societies choose to fail or succeed. Viking, New York, 592 pp

Jacobsen H-A (2013) Energy informatics: opportunities and challenges in interdisciplinary research. Presentation of the middleware systems research group of the Technische Universität München

Nash J (1951) Non-cooperative games. Ann Math 54(2):286–295

NIST (2014) Smart grid framework 3.0

Prigogine I, Stengers I (1984) Order out of Chaos: man's new dialogue with nature. Flamingo, London

Williams B (1985) Ethics and the limits of philosophy. Fontana Press, London

References of Section 2.2

Mager U (2010) Die Entwicklung des Wasserwirtschaftsrechts - Referenzgebiet für ein materiell-rechtlich fundiertes internationales Verwaltungsrecht [The development of water management law – a reference field for international administrative law based on substantive law]. In: ZaöR V (2010) 70:789–818

Schuler H, Prochaska M, Frintrup A (2001) Leistungsmotivationsinventar [Achievement motivation inventory] (LMI). Hochgrefe Verlag, Göttingen

Vogt M (2009) Prinzip Nachhaltigkeit - Ein Entwurf aus theologisch-ethischer Perspektive [The principle of sustainability – a draft from a theological-ethical perspective]. ISBN 978-3-86581-091-5

Zippelius R (2011) Rechtsphilosophie [Legal philosophy], vol 6. Beck-Verlag, München. ISBN 978-3-40661191-9

References of Section 2.3

Bundesnetzagentur (2011) „Smart Grid" und „Smart Market"- Eckpunktepapier der Bundesnetzagentur zu den Aspekten des sich verändernden Energieversorgungssystems. Bundesnetzagentur für Elektrizität, Gas, Telekommunikation, Post und Eisenbahnen

Bundesnetzagentur (2014) Monitoring-Bericht 2014. Bundesnetzagentur für Elektrizität, Gas, Telekommunikation, Post und Eisenbahnen

Bundesumweltministerium (2013) Erneuerbare Energien in Zahlen Nationale und internationale Entwicklung. Bundesministerium für Umwelt, Naturschutz und Reaktorischerheit (BMU)

Faulstich M (2012) Energie und Wirtschaft in Deutschland: Gedanken zur Energiewende. Faszination Forschung (10). Technische Universtität München

Geisberger E, Broy M (eds) (2012) agendaCPS – Integrierte Forschungsagenda Cyber-Physical Systems. *acatech STUDIE*. Springer, Heidelberg

Kießling B (2013) Design and evaluation of a hierarchic ICT-architecture for the integration of intelligent prosumers into a smart grid. Technische Universität München

Koss D, Bytschkow D, Gupta P, Schätz B, Sellmayr F, Bauereiß S (2012) Establishing a smart grid node architecture and demonstrator in an office environment using the SOA approach. In: Software engineering for the smart grid

Piaszeck S, Wenzel L, Wolf A (2013) Regional diversity in the costs of electricity outages: results for German counties. Hamburg Institute of International Economics (HWWI)

Schätz B (2014) The role of models in engineering of cyber-physical systems – challenges and possibilities. CPSWeek 2014

Tanenbaum A Wetherall D (2013) Computer networks. Pearson International, London, UK

Reference of Section 2.4

Bontrup H-J, Marquardt RM (2015) Die Zukunft der großen Energieversorger, Hannover, Lüdinghausen, http://www.greenpeace.de/sites/www.greenpeace.de/files/publications/zukunft-energieversorgung-studie-20150309.pdf

Breyer C et al (2014) Vergleich und Optimierung von zentral und dezentral orientierten Ausbaupfaden zu einer Stromversorgung aus Erneuerbaren Energien in Deutschland, Reiner Lemoine Institut gGmbH, Haleakala-Stiftung, Berlin

Diamond JM (2005) Collapse. How societies choose to fail or succeed. Viking, New York

Felber C (2010) Gemeinwohl-Ökonomie. Das Wirtschaftsmodell der Zukunft, Wien

Hill RA, Dunbar RI (2003) Social network size in humans. Hum Nat 14(1):53–72

Holling CS (1995) What barriers? What bridges? In: Light SS (ed) Barriers and bridges to the renewal of ecosystems and institutions. Columbia University Press, New York

Holling CS (2001) Understanding the complexity of economic, ecological and social systems. Ecosystems 4(5):390–405

Holling CS, Gunderson LH, Ludwig D (2002) In quest of a theory of adaptive change. In: Holling CS (ed) Panarchy. Island Press, Washington, DC

Paech N (2012) Befreiung vom Überfluss. Auf dem Weg in die Postwachstumsökonomie. Oekom Verlag, München

Plankl R (2013) Regionale Verteilungswirkungen durch das Vergütungs- und Umlagesystem des Erneuerbare-Energien-Gesetzes (EEG), working paper, Thünen-Institut für Ländliche Räume, Braunschweig

Stiglitz JE (2003) Globalization and its discontents. WW Norton & Company, New York

Chapter 3
Governance of Societies

Christian Werthmann, Ortwin Renn, Anastassia Makarieva, Victor Gorshkov, Peter A. Wilderer, Verena Risse, Carolin Böker, and Bettina Haas

3.1 Skipping Centralization

Christian Werthmann

C. Werthmann (✉); Corresponding author for Section 3.1
Leibniz University, Hannover, Germany
e-mail: werthmann@ila.uni-hannover.de

O. Renn (✉); Corresponding author for Section 3.2
University of Stuttgart, Stuttgart, Germany
e-mail: ortwin.renn@soz.uni-stuttgart.de

A. Makarieva (✉); Corresponding author for Section 3.3 • V. Gorshkov
National Research Centre "Kurchatov Institute" Petersburg Nuclear Physics Institute, St. Petersburg, Russia
e-mail: elba@peterlink.ru

P.A. Wilderer (✉); Corresponding author for Section 3.3
Institute for Earth System Preservation (IESP), Technical University of Munich, Garching, Germany
e-mail: peter.wilderer@mytum.de

V. Risse (✉); Corresponding author for Section 3.4
Munich School of Philosophy and Frankfurt University, Munich, Germany
e-mail: Verena.Risse@hfph.de

C. Böker (✉); Corresponding author for Section 3.5
Technical University of Munich, Munich, Germany
e-mail: carolin.boeker@tum.de

B. Haas
Bavarian State Ministry of the Environment and Consumer Protection, Section on Soil Protection, Contaminated Sites and Geology, Munich, Germany

P.A. Wilderer, M. Grambow (eds.), *Global Stability through Decentralization?*,
Strategies for Sustainability, DOI 10.1007/978-3-319-24358-0_3

3.1.1 Key Messages

It is no secret that the world is urbanizing very fast. If UN estimates turn out to be true, by midcentury 2.5 billion people will have joined the approximately 3.9 billion living in cities today, an urban population increase of over 60 %! It is one of the biggest questions of our time, how the 6.4 billion who will live in cities by 2050 will be fed, housed and provided with basic services in a sustainable way. What kind of infrastructure and governance models will be the most healthy in the progressing urbanization of the planet? What is the best mix between decentralization and centralization of services, infrastructure and government structures?

These are highly complex questions and there are no simple answers to complex questions especially in complex contexts. Urbanization is diverse, living is diverse, situations differ, landscapes differ, cities vary, people vary. As a trained landscape architect, I would say, the answers to the above questions are highly dependent on the context, on the social, physical and political landscape they are embedded in.

3.1.2 Where and How Urbanization Will Occur

Future urbanization processes on the planet will be very unevenly distributed. Most of our global urban growth in the coming decades will occur in Asia and Africa. Much of South America is already urbanized (for instance, 90 % in Brazil), North American cities will only grow moderately, and many European cities will rather stagnate in population numbers unless immigration policies will change (some do even shrink). That means that Europe and large parts of the Americas are experiencing urban consolidation processes where existing cities will be redeveloped and restructured (hopefully in a more sustainable way), while cities in Africa and Asia will rapidly expand.

What will this coming urbanization look like? Urbanization processes will look very differently pending on the respective development level of a singular country, region or city. In more authoritarian governments with centralized planning agencies the development of new cities and urban additions will be centrally planned in a strict top down process. For example the Egyptian government plans to build 40 new cities to house millions in the desert. China plans to build homes for an additional 100 million inhabitants by 2020. Brazil has programs to build millions of new homes under a federal program called Minha Vida, Minha Casa.

Vast master plans are drawn up covering many square kilometers with the attempt to create new living environments for millions from scratch. As previous new town examples have shown, the success and livability of formally planned city expansions are questionable; authoritarian master-planned cities with little to no participation processes rarely function as desired. They often lack employment opportunities, public transportation, affordability, social cohesion and diversity. There are not many examples in the world where the creation of new towns has been completely successful. The creation of new homes and livelihoods is a highly complex undertaking and one could question how much of it can be even planned.

It would go beyond the frame of this essay to go into further detail, especially because formally planned urban expansions by governments with sufficient funding capacities will be actually the smaller part of planetary urbanization.

Much of the urbanization pressure will be in the developing and least developed countries of Africa and Asia where urban planning, funding and implementation capacities are limited. Per UN Habitat, the majority of future urban growth will occur in cities with less than 500,000 inhabitants, throwing overboard common perceptions of an urban world dominated by sprawling megacities (cities over 10 million inhabitants). These smaller to mid-sized cities will double, triple, quadruple in size and are typically not very well equipped nor have the financial resources to handle this type of massive increase of their populations. In the absence of viable alternatives, the majority of these urban expansions will occur outside of legal planning frameworks (if they exist). People will build houses and neighborhoods if not whole cities on their own without a master plan or official city permit. These people are forced to act on their own, not because they act like criminals, but because they have no other choice (lack of serviced land, lack of affordability, common custom). If the UN prognosis is correct, the majority of our future urbanization will be dominated by what UN-Habitats coins as "slums". Slums are not always correlated with informal housing activity, but most informal neighborhoods in their early stages would fall under this definition. UN-Habitat estimates that almost one billion people currently live in slum-like conditions. It is estimated that by 2050 two billion more will live in slum-like conditions, or two-thirds (!) of our new urban population. That means we are looking at a global urbanization process where city planning as a classical achievement of normative centralization efforts will be the exception and decentralized decision making as a ancient act of settlement formation will be the rule.

Decentralization as the Urban Default

De facto, the majority of our future urbanization will occur in a predominantly decentralized manner where many individuals will make individual decisions where to build a house or an annex thereby creating completely new cities. These individuals will make these decisions based on a complex set of aspects with the proximity to employment as the overriding factor. Most of the new informal urban expansions will have very little public infrastructure as in cities of the developing world basic accommodations like running water, permanent electricity or paved roads (a given in cities of highly industrialized countries) can be more the exception than the rule. In rapidly industrializing countries with growing prosperity and democratic governments (Brazil, Chile, Colombia, Argentina etc) informal settlements might gain some form of access to water, sanitation, public transport, schools or health services once they have been accepted by authorities and are not anymore in danger to be forcibly removed. There are many examples especially in Brazil, Colombia and other parts of Latin America where informal settlements have been retrofitted with new social and technical infrastructure. They are typically connected to a centralized infrastructure system (water, sewage, electricity) based on nineteenth and twentieth century technologies that these countries have inherited from more industrialized countries. Centralized infrastructure emblematic through flush toilets or central heating is seen as a sign of modernity, progress and urbanity; decentralized infrastructure systems like rain water harvesting or local energy production can be valued as signs of a rural lifestyle that most of the migrants tried to leave behind by moving to the city.

That means we live in some kind of a time warp. While highly industrialized countries like Germany think about more decentralizing their hard-won twentieth century achievements of centralized services like sewage treatment, potable water provision, energy and food production, less industrialized and industrializing countries are going full speed for centralized services in their cities.

3.1.3 *The Case of São Paulo*

How centralized infrastructure and large scale engineering projects have problems to manage the complex metabolism of a megacity can be studied on the case of São Paulo and its water management. São Paulo grew in the last century from a mid-sized town to a 20 million metropolis. Over that time it adopted and implemented the centralized engineering and infrastructure approaches typical for industrialized countries. São Paulo's water management is indicative of a classical sectoral thinking where potable water provision, stormwater management and sewage treatment is thought of as separate problems and handled through separate large-scale infrastructure systems. Potable water is provided through a series of reservoirs far away from the city. Older reservoirs closer to the city are engulfed by urbanization and are only in small parts usable. Most lately (2015) São Paulo is facing the largest water shortage in its recent history, with reservoirs almost depleted, whole neighborhoods without running water and no additional water sources available in the short term (there are plans to build new reservoirs in the North).

While São Paulo is pondering how to master its water shortage, it releases its precious used water into the rivers flowing away from the city. Sewage treatment is a fairly new concept and São Paulo has built several large treatment plants only in the last decades. The system is characterized by a few plants that process the sewage of millions, often pumping effluent for over 70 km across neighboring watersheds. On top of the great energy loss of pumping, approximately one third of the water is lost on its way through leaky pipes. The remainder of the treated sewage is then finally released into the rivers eventually flowing thousands of kilometers later into the Atlantic Ocean.

Water scarcity has not been on the list for the typical problems of São Paulo. To think about a decentralized recycling and harvesting water system in the city itself is a fairly new consideration that São Paulo had not to make so far in a tropical climate with lots of rainfall. Normally it had the opposite problem. São Paulo is known for its crippling floods. The city is built over an inner delta where several rivers confluence. All rivers are channeled, all floodplains have been lost to urbanization and all major highways and railway lines run along the rivers. Given the high impermeability of the urbanized watershed, the engineering of the channels cannot keep up with the amounts of water flowing through, hence flooding occurs. The next flood will come for sure, turning the cities attention away from the drought problem.

Floods, droughts, pollution—it is clear that a more integrative decentralized management of water is needed where water is harvested, treated and recycled in its respective watershed. In contrast to the twentieth century, São Paulo has now moderate growth rates and the city has the chance to reconstitute itself and hopefully will rethink its approach towards a more integrative infrastructural thinking. It is not only the technical issues that are insufficient, its stream channeling and subsequent highway construction have robbed the city from its fluvial landscape as a fundamental recreational asset that will be very hard to regain.

3.1.4 Skipping Centralization

São Paulo is special in its size and financial might, and as mentioned before most urban growth on the planet will occur in cities below 500,000 inhabitants in locations with less financial resources. However, São Paulo can still provide important lessons for cities which are right now experiencing the frantic growth spurts that São Paulo had to register in the twentieth century. For these mid-sized cities it might be worthwhile to try an alternate approach of how to service their new citizens with basic infrastructure that is more adaptive and resilient. There is a historical chance to leapfrog traditional, sectoral large scale infrastructure provision towards a more decentralized, approach that begins at the individual household, extends to the block, neighborhood, district and eventually to the city and its region, and where at each scale the appropriate level of service provision is handled.

Thereby it will not be a matter of just transferring decentralized infrastructure systems of highly industrialized countries such as green roofs, solar panels, small scale power plants, sewage treatment wetlands or rooftop farming to less industrialized countries. As laid out before, two thirds of our future global urbanization will be informal, an urbanization typology that is at its core a radically decentralized decision apparatus. It will be of matter that the new, mostly low income residents will be supplied with the capacity to make sustainable decisions about their environment versus being connected after the fact to a resource wasting apparatus that brings small cities into financial turmoil. One could argue for a hybrid approach where cities with little funding capacity get the help to anticipate the incoming migrants and actively engage them in the planning and building process of their future environments. Historic ideas of the 1970s to provide land and core housing are currently being revived and need to be updated towards a more participatory and integrative public space and infrastructure design. More decentralized, localized infrastructure could be a source of jobs and revenue that is desperately needed by low-income residents.

There are substantial challenges to implement a more livable and sustainable environment for future urban populations. Unavailability of land, unclarity of land ownership, real estate speculation, negative attitudes towards newcomers and new climate induced hardships are just a few among them. The hardest challenge in my eyes though will be to overcome the lure of twentieth century Western technology that magically brings running water and electricity into everybody's house without exposing the side effects of urban degradation and high utility bills.

3.2 The Balance Between Efficiency, Effectiveness, Resilience and Cohesion

Ortwin Renn

3.2.1 Key Messages

The design of policy making should be guided by a discursive attempt to find the optimal balance of all four sectors of society, namely effectiveness, efficiency, resilience and social cohesion. This may also imply that in some instances highly centralized systems should be preferred while in other instances decentralized solutions provide the better alternative.

3.2.2 Basic Considerations

At the foundation of sustainable development is the need for a well-rounded balance between effectiveness, efficiency, resilience and social cohesion.

Effectiveness refers to the need of societies to have a certain degree of confidence that human activities and actions will actually result in the consequences that the actors intended when performing them.

Efficiency describes the degree to which scarce resources are used to reach the intended goal. The more resources are invested to reach a given objective, the less efficient the activity under question remains.

Resilience describes the capacity to sustain functionality of a system or a service even under severe stress or unfamiliar conditions.

Finally, *social cohesion* covers the need for social integration and collective identity despite plural values and lifestyles.

All four needs or functions of society build the foundation for legitimacy. Legitimacy is a composite term that denotes, first, the normative right of a decision-making body to impose a decision even on those who were not part of the decision-making process (issuing collectively binding decisions), and second, the factual acceptance of this right by those who might be affected by the decision. As a result, it includes an objective normative element, such as legality or due process, and a subjective judgment, such as the perception of acceptability.

Within the macro-organization of modern societies, these four functions are predominantly handled by different societal systems: economy, science (expertise), politics (including legal systems), and the social sphere. In the recent literature on governance, the political system is often associated with the rationale of hierarchical and bureaucratic reasoning; the economic system with monetary incentives and individual rewards; and the social sphere with the deregulated interactions of groups within the framework of a civil society (Rosa et al. 2014; Renn 2014; Parsons and Shils 1951) Another way to phrase these differences is by distinguishing among competition (market system), hierarchy (political system), and cooperation (sociocultural system).

Each of the four systems can be characterized by several governance processes and structures adapted to the system properties and functions in question.

In the market system, decisions are based on the cost-benefit balance established on the basis of individual preferences, property rights, and individual willingness to pay. The conflict resolution mechanisms relate to civil law regulating contractual commitments, Pareto optimality (each transaction should make at least one party better off without harming third parties), and the application of the Kaldor–Hicks criterion (if a third party is harmed by a transaction, this party should receive financial or in-kind compensation to such an extent that the utility gained through the compensation is at least equivalent to the disutility experienced or suffered by the transaction). The third party should hence be at least indifferent between the situation before and after the transaction. In economic theory, the transaction is justified if the sum of the compensation is lower than the surplus that the parties could gain as a result of the planned transaction. However, the compensation does not need to be paid to the third party. Additional instruments for dealing with conflicts are (shadow) price setting, the transfer of rights of ownership for public or non-rival goods, and financial compensation (damages and insurance) to individuals whose utilities have been reduced by the activities of others. The main goal here is to be efficient. In most instances, up-scaling and centralizing production and distribution services improve the scores for efficiency.

In politics, decisions are made on the basis of institutionalized procedures of decision-making and norm control (within the framework of a given political culture and system of government). The conflict resolution mechanism in this sector rests on due process and procedural rules that ideally reflect a consensus of the entire population. In particular, decisions should reflect the common good and the sustainability of vital functions to society. This is why resilience lies at the heart of public activities. In democratic societies, the division in legislative, executive, and judicial branches; defined voting procedures; and a structured process of checks and balances underscore the institutional arrangements for collective decision making. Votes in a parliament are as much a part of this governance model as is the challenging of decisions before a court. The target goal here is to seek resilience as a major prerequisite of legitimacy. Both resilience and legitimacy are best served by decentralized systems as they provide diversity, more control options and less dramatic effects if one system fails.

Science has at its disposal methodological rules for generating, challenging, and testing knowledge claims, with the help of which one can assess decision options according to their likely consequences and side effects. If knowledge claims are contested and conflicts arise about the validity of the various claims, scientific communities make use of a wide variety of knowledge-based decision methods, such as methodological review or re-tests, meta-analysis, consensus conferences, Delphi, or (most relevant in this arena) peer review to resolve the conflicts and test the explanatory or predictive power of the truth claims. These insights help policymakers understand phenomena and be effective in designing policies. To be effective is not related to the degree of centralization. However, effectiveness can be an important moderator between efficiency and resilience.

Finally, in the social system, there is a communicative exchange of interests, preferences, and arguments assisting all actors to arrive at a unanimous solution.

Conflicts within the social system are normally resolved by finding favorable arrangements for all parties involved, using empathy as a guide to explore mutually acceptable solutions, referring to mutually shared beliefs, convictions, or values or relying on social status to justify one's authority. These mechanisms create social and cultural cohesion. The most important aspect here is fairness towards the present and the future generation. Fair solutions tend to be more decentralized but often at the expense of optimal allocation of resources and services.

Socially relevant problems are rarely dealt with within the limits of one single system rationale. Instead, they go through interrelated procedures, either sequentially or in parallel. For example, the political system can decide on a specific goal or target by parliamentary vote (e.g., a limit on automobile emissions) and then leave it to the market to implement this decision (such as organizing an auction to sell emission rights to all potential emitters). Or a governmental decree is reviewed by an expert panel or a citizen advisory committee. Of particular interest are decision-making processes that combine the logic of two or more systems. The settlement of conflicts with the method of mediation or negotiated rulemaking can, for example, be interpreted as a fusion of economic and social rationale. The cooperation between experts and political representatives in joint advisory committees (i.e., the experts provide background knowledge, while politicians highlight preferences for making the appropriate choices) represents a combination of knowledge-oriented elements and political governance. Classic hearings are combinations of expert knowledge, political resolutions, and the inclusion of citizens in this process.

3.2.3 Conclusion

These insights suggest that for complex policy decisions that are crafted to enhance the sustainability of society representatives of all four sectors of society need to be included in order to ensure that decisions are effective, efficient, resilient and fair. It seems also prudent to conclude that representatives of one sector should not be able to outvote the representatives of the others sectors since each contribution is needed for sustainable decision making. Maximizing efficiency on the expense of the other goals may compromise resilience and maximizing effectiveness may compromise fairness.

3.3 What Can We Learn from Natural Ecosystems to Avoid a Civilization Breakdown?

Anastassia Makarieva, Victor Gorshkov, and Peter A. Wilderer

3.3.1 Key Messages

An attempt is made to formulate a comprehensive cross-disciplinary view on the environmental problems of modern humanity by considering principles of order maintenance in living systems at different levels of organization. It is argued that while decentralization is key to order maintenance in living systems, which occurs by competitive interaction of independent units, the ecological services provided by the biota in terms of climate and environmental stabilization range in scale from local to regional and global. Thus, international and global cooperation is indispensable for preserving these services. The role of large animals in ecosystem stability is discussed together with its implications for the ecological requirements of *Homo sapiens*. It is suggested that adding an ecological dimension to the conventional hierarchy of human needs and motivations can shed light on many important problems of modern society.

3.3.2 Introduction: Does Civilization Progress Have a Direction?

During the last two centuries of rapid scientific and technological progress an advanced set of views emerged on the relationship between the *Homo sapiens* and the biosphere. The accumulated data on biological fossils testified that in the course of the biological evolution morphology and behavior of species were getting more and more complex and sophisticated. It appears that *Homo sapiens* is—at least at the time being—the winner of the evolutionary process. Owing to the scientific and technological progress humans have colonized practically all land having displaced other species from their natural ranges. The scientific and technological progress is considered—in analogy to evolution—as a process during which the human society has gotten progressively more and more complex and organized. As a result, the ever-increasing complexity is expected to trigger an increase of energy consumption since a diminishing flow of external energy is known to drive all systems to a state of thermodynamic chaos. At the same time, it is commonly assumed that the ongoing transformation of natural ecosystems is not a major threat for our global civilization. Despite the fact that many natural species of the biosphere are already eradicated and replaced by artificially modified sorts of plants and animal breeds it is generally believed that human life is not in danger.

At the commencement of the era of "Internet of Things", in Germany called "Industry 4.0 (forth industrial revolution) (Aslak and Bruaset 2013; Dombrowsky and Wagner 2014), this spontaneous development continues even though the currently available knowledge does not allow a solid evaluation of its consequences. But there are efforts to make this development sustainable by, for instance, keeping economic growth from getting undermined by either political or ecological crises. However, unlike the sustainable development of an embryo that transforms into an adult governed by the genetic program of its species, the scientific and technological progress as well as the economic development is not governed by such a program. Sustained movement of the global civilization in an unpredictable direction can lead to a global catastrophe. To examine and exclude unfavorable scenarios it is needed to elaborate a scientific theory that would be able to predict the future of the civilization based on the known laws of nature and the scientific evidence about the human society and the biosphere. In this article we make an attempt to advance in this direction.

Life on Earth is at least 3.8 billion years old (Mojzsis et al. 1996). In the face of the myriad external forcings that impacted our planet during the last four billion years the process of life never discontinued. It is therefore of interest what the major principles are that underlie such an apparently unique sustainability and resilience. In the first part of this paper we attempt to outline such principles to show that the maintenance of order in life is essentially decentralized and strongly contingent upon interactions between numerous and independent living systems at different levels of complexity. We then discuss how living order maintenance is inseparable from environmental sustainability, which thus emerges as a product of life functioning. Thus a global scale loss of natural ecosystems is incompatible with long-term environmental safety.

By analyzing the energy flow through natural ecological communities we discuss how the environmental impact of different living organisms crucially depends on their body size. Large animals including humans are potential destabilizers of ecosystem biomass and productivity and have very specific functions in natural ecological communities.

In the second part of the paper we consider how scientific knowledge about life-environment interaction and principles of ecological sustainability can enhance understanding of the global problems faced today by our civilization. We discuss how considerations of the ecological requirements of *Homo sapiens*—traditionally excluded from consideration of the pyramid of human needs (e.g., Kenrick et al. 2010)—can provide clues to understanding important global processes of today.

Indeed, while there are many optimistic voices about a bright future associated with further technological progress, there are also well-substantiated doubts spreading as to whether such progress is still able to further improve human conditions as it used to do (Cowen 2011; Atkinson and Ezell 2012; Mokyr 2013; Gordon 2014).

With presenting a broad picture of seemingly diverse but, as we argue, deeply interrelated concepts and problems we aim to stimulate a discussion of how the evidence from different fields of modern science—from ecology and genetics to climatology and economics—could be meaningfully synthesized across disciplines into a coherent scientific framework that would guide the development of our civilization and allow us to avoid a global collapse.

3.3.3 Informational Precipice Between the Biosphere and Civilization

One apparent feature of life is its complexity in comparison with the inanimate world. We begin our consideration by quantifying this complexity in terms of information fluxes operated by living systems.

The Sun sends to the Earth ordered energy in the form of short-wave photons. On Earth this energy transforms into the chaotic energy of thermal photons that are emitted back to space. Photon's energy is proportional to kT, where $k = 1.4 \times 10^{-23}$ J K^{-1} is Boltzmann constant, T is the absolute temperature of radiation. The energy of solar and thermal photons is determined by temperatures of the Sun and the Earth, respectively, $T_S \sim 6000$ K and $T_E \sim 300$ K. The energy of solar photon as it dissipates into thermal photons is conserved. Hence, we have $kT_S = n\ kT_E$ and $n \approx 20$: each solar photon decays on Earth into about twenty thermal photons.

This decay can go through different channels. The unbounded diversity of possible decay channels maintains all ordered processes on Earth, both in animate and inanimate nature. If the Sun were sending to the Earth the same flux of energy but in the form of thermal photons that are emitted by the Earth, the temperature of the Earth's surface could be about the same as it is now. But the decay of these thermal photons would be impossible. All the decay channels were closed. The Earth would remain warm, but no ordered processes would occur on its surface. Life could not exist (Fig. 3.1).

The main difference between life and inanimate nature pertains to the fact that life uses decay channels that are by many orders of magnitude more numerous and complex than in the non-living world. Orderliness of biological systems is characterized by molecular (not macroscopic as in the inanimate nature) "memory cells" or degrees of freedom. Per each square micron of the Earth's surface there are

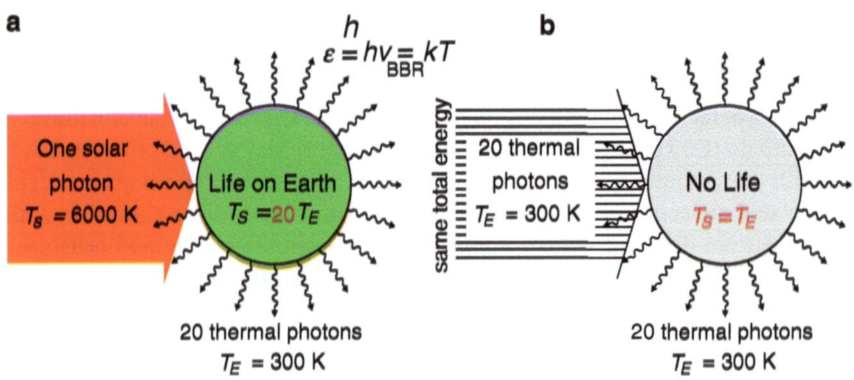

Fig. 3.1 Solar energy transformation on Earth. (**a**) 20 thermal photons $T_B = 300$ K. (**b**) 20 thermal photons $T_B = 300$ K

several independently functioning living cells—plankton in the ocean, plants, bacteria and fungi on land. These cells react to local changes of their environment in a non-random way. They exchange energy, matter and information with the environment as prescribed by their genetic program coded in the DNA molecules.

Radiation consists of particles—photons. Photon energy is $\varepsilon = h\nu = kT$ for blackbody radiation (BBR) of temperature T. Radiation power is $J = n\varepsilon$ (n is the number of photons radiated per unit time). (a) Earth receives power $J_S = 1.4 \times 10^{17}$ W of solar radiation with $T_S \approx 6000$ K and emits power $J_E = J_S$ of thermal radiation with $T_E \approx 300$ K. Energy on Earth does not accumulate (steady state): $J_S = J_E$, so $n_E/n_S = \varepsilon_S/\varepsilon_E = T_S/T_E = 20$. Each solar photon decays on Earth into 20 thermal photons. Any events on Earth are possible because Earth's temperature is 20 times less than Sun's. (b) If those temperatures coincided, Earth would have been almost completely "uneventful", existing in a state close to thermodynamic chaos. Life on Earth would be impossible even if the Earth remained as warm as it is now.

The rate of information exchange between the living cells and their environment can be estimated from the known rate of their energy consumption. Absorption of one solar photon by a plant cell changes the state of about twenty molecular memory cells within the cell as the solar photon decays into thermal photons. Assuming that one molecule corresponds to one memory cell with two possible states—excited (after absorption of energy of the order of kT_E) and non-excited (after release of this energy)—we obtain that one act of excitation and relaxation corresponds to a flux of information of one bit per act. With the global mean efficiency of photosynthesis of about $\varepsilon = 0.5$ %, global mean flux of solar energy absorbed by the planetary surface of about $F = 170$ W m^{-2}, total global flux I of information processed by living cells on the Earth's surface of area $S_E = 5 \times 10^{14}$ m^2 is estimated as $I = \varepsilon F S_E/(kT_E) = 10^{35}$ bit s^{-1}.

There is virtually a precipice between the information processing capacities of the biosphere and our civilization. It pertains to the total fluxes of information as well as the energy efficiency of information processing. If all people on Earth had a modern PC that runs about 10^{11} operations per second, total flux of information processing by the humanity would not exceed 10^{21} operations per second, which is 14 orders of magnitude less than in the biosphere. Real rates of information processing in our civilization are much lower. For example, GOOGLE search processes data at rate of about 10^{13} bit s^{-1}, i.e. by 22 orders of magnitude slower than does the biosphere (Dean and Ghemawat 2008).

Modern supercomputers are able to perform about 10^{16} operations per second, occupy an area of about 10^2 m^2 and consume power of about 10^7 W. Their energy expenditure per operation—about 10^{-9} J per operation—is 12 orders of magnitude larger than in the biosphere ($kT_E \approx 4 \times 10^{-21}$ J). If the entire Earth's surface had been covered with such supercomputers their total flux of information processing would be 5×10^{28} bit s^{-1}. This is two million times less than in the biosphere. Meanwhile the energy consumption rate of such a global computer network would have been 500 times larger than the flux of solar energy at the surface, a hundred

Fig. 3.2 Technological and ecological information processing systems. *Left panel*: Tianhe-2, world's fastest supercomputer located in China (Photo credit Jack Dongarra). Tianhe-2 performs 34×10^{15} operations per second, occupies 720 m^2 and consumes 24×10^6 W. The supercomputer information processing rate per unit area is 5×10^{13} operations per second per square meter and energy consumption is 3×10^4 W per square meter. *Right panel*: Rainforest in Papua New Guinea (Photo credit Rocky Roe and UPNG Remote Sensing Centre) has a rate of information processing 2×10^{20} operations per second per square meter, which is over a million times faster than Tianhe-2. The rainforest energy consumption does not exceed the solar power flux of 200 W m^{-2}, i.e. the forest is at least a hundred times more efficient than Tianhe-2

thousand (10^5) times larger than the energy consumption of the biosphere and one million times larger than the energy consumption of modern civilization (Fig. 3.2).

To make use of the huge diversity of possible decay channels for solar photons life must have been minimized energy losses within each channel. The energy efficiency achieved by life is unprecedented. An egg transforms to a chicken without any external energy consumption. Internal energy losses (heat dissipation) during embryonic development in some reptilian and insects do not exceed 10 % of the initial energy store of the egg (Makarieva et al. 2004a). The egg-to-chicken transformation represents an irreversible process of decay that is characterized by a diversity of channels that is unimaginable in the civilization. Living and non-living nature consume not energy but the information of the Sun. Living matter uses this information with maximum efficiency.

3.3.4 Biotic Regulation of the Environment

Thus, we can view the biosphere as a global distributed network of microscopic computers. Total number of such simultaneously working computers (living cells) in the biosphere is in the order of 10^{28}–10^{30} (Gorshkov 1995; Whitman et al. 1998). From the same perspective life can be viewed as a unique self-sustainable algorithm that has been operating on Earth governed by the genetic program of the living cells for about four billion years. One can ask: what information must be contained in the genetic program of life to account for such an extraordinary persistence?

Apparently, information about copying (reproduction) of living objects is an indispensable but also the simplest module of that program. Indeed, copying is common to many simple processes in the inanimate nature—for example, to chain reactions. What is unique about life is that the algorithm by which the living organisms re-create themselves has never aborted during the nearly four billion years of life existence. Since life as a whole can only exist in a narrow interval of external conditions, life's persistence means that the genetic program of life comprises information about what the suitable for life conditions are and how to maintain them. If this information changes in the course of biological evolution, this entails respective changes in the environment maintained by life. Thus, environmental changes by themselves do not necessarily represent evidence of broken biotic control.

The suitable for life environment is physically unstable. Biota (the totality of natural living organisms) uses the huge information fluxes to control the environment and stabilize it in an optimal for life state. Below we briefly discuss several key aspects of the biotic regulation of the environment (Gorshkov 1995; Gorshkov et al. 2000).

For the biota to function, stores of organic as well as inorganic carbon must be present in the environment. These stores in the modern biosphere are of the order of 10^3 Gt C (1 Gt $= 10^9$ t). For example, atmospheric carbon dioxide, which is necessary for the photosynthesis, contains 700 Gt C. The biotically active stores of organic carbon in soil, wood and ocean are of the same order of magnitude. The biota recycles carbon at a mean global rate of about 10^2 Gt C/year. If the fluxes of synthesis and decomposition of organic carbon by the biota had not been strictly correlated with each other (as they are not correlated, for example, in our civilization), the stores of carbon in either organic or inorganic reservoirs would have been depleted just in several decades. To stabilize the reservoirs of all life-important elements is only possible if the organisms that synthesize and decompose organic matter interact with each other in a non-random, coordinated manner such that in the course of this interaction all deviations of the environment from an optimal state are compensated.

A conspicuous example of the biotic control of global environmental conditions on a geological scale is the atmospheric carbon. Carbon dioxide enters the atmosphere in the result of magmatic and metamorphic degassing in geological processes related to volcanoes, continental rifts, seismically active regions etc. The removal of atmospheric CO_2 in inorganic form occurs by weathering: the formation of carbonate rocks from silicate rocks (Berner 1990). These opposing processes are controlled by independent factors: e.g., weathering strongly depends on the size and elevation of the continents as well as on the intensity of the river runoff. The rates of inorganic carbon removal and burial are such that any of these processes alone could have changed the atmospheric CO_2 concentration by an order of magnitude in just a few thousand years (Berner 1990). While the biota profoundly impacts weathering, i.e. the rate of carbonates formation (Berner 1990; Schwartzman and Volk 1989), the two opposing fluxes of the inorganic carbon do not match. The net rate of carbon emission to the atmosphere turns out to be positive and similar in the order of magnitude to the gross inorganic carbon fluxes by weathering and degassing (Garrels and Lerman 1981). On a longer time scale—over Phanerozoic time—such an imbalance between weathering and degassing could have brought about catastrophic fluctuations of the atmospheric carbon concentration. However, no catastrophic changes in atmospheric CO_2 amounts actually took place as the biota deposited the excessive carbon in sediments in the form of inert organic compounds. If the carbon deposited in organic form (about 10^7 Gt C) had remained in the atmosphere, CO_2 concentration would have been a hundred thousand times higher than it is now (Gorshkov 1995; Gorshkov et al. 2000).

In the discussions of life-environment interactions the focus has conventionally been on *recycling*: the life-mediated enhancement of the geological fluxes of elements (e.g., Downing and Zvirinsky 1999; Free and Barton 2007). The carbon cycle example discussed above illustrates that the main principle of biotic regulation is not a *recycling* of life-important chemical elements (see Chap. 20, Bloesch) but a directional compensation of their unfavorable environmental changes. It is not closeness, but a non-random *openness* of the biochemical cycles that result in environmental homeostasis. Without this mechanism the uncorrelated fluxes of inorganic substances to and from the biosphere would have made it unsuitable for life in a relatively short period of time.

Photosynthesis is the energetic basis of modern life. For it to be possible, temperature of the Earth's surface should be compatible with the liquid phase of water, i.e., it must be higher than 0 °C. On the other hand, it cannot exceed or approach ~60 °C, which is the limit when cell structures start to disintegrate. Only very few species of archae and bacteria, termed extremophiles, can live at ambient temperatures of even higher than 100 °C, but they do not photosynthesize (Anitori 2012; Canganella 2012). Meanwhile for the Earth's surface, two thirds of which are covered by the ocean, two physically stable states are a completely glaciated Earth with surface temperature of about −100 °C and an Earth with its oceans evaporated and surface temperature about +400 °C. In the absence of stabilizing biotic processes a random climate state that occasionally happens to be suitable for life would undergo transitions to any of the two stable states in time periods of the order of thousand years. Biotic regulation of the environment has ensured biotic stability of the environment with a global mean temperature in the vicinity of 15 °C over the entire period of life existence (Makarieva and Gorshkov 2001; Gorshkov and Makarieva 2002).

Despite occupying over two thirds of the Earth's surface, global biological productivity of the ocean is smaller than productivity of forests and swamps on land. Since the forest cover formed on land about three hundred million years ago land biota has been playing a major role in the regulation of the global environment and climate. Evolution of tree plants made it possible for the biotic pump of atmospheric moisture (Makarieva and Gorshkov 2007; Sheil and Murdiyarso 2009; Makarieva et al. 2014) to operate on land, which enabled life to colonize all land. Moisture evaporated from the ocean surface is transported to the continental interior only in the presence of an extensive forest cover. Forest cover absent, land can turn into a lifeless desert on a time scale of a few decades. Thus, water on land, which is indispensable for human life, is also controlled by the biota (Makarieva and Gorshkov 2010; Wilderer 2009).

3.3.5 Principles of Order Maintenance in Life: Decentralization Is Key

Theoretical biology conventionally highlights evolution as the central process of life, rather literally reflecting the widely quoted formulation of "nothing in biology making sense except in the light of evolution" (Dobzhansky 1973). Scientists debate how and why the genetic information of species changes with time. A more fundamental question, however, receives little if any attention: why no erosion of life information has occurred in four billion years. With mutation rates in the DNA-based world being universal, about 10^{-10} base pairs per division, and assuming about 10 divisions in the germ line as a grand mean for life based on its universal metabolic rhythm (Makarieva et al. 2008), we conclude that all genetic information of life could have been completely eroded by mutations in one billion years. In other words, the genomes of species would have represented a chaotic random sequence of the four genetic letters (base pairs). Since about 1 % of the genome does not tolerate any changes at all (such changes are incompatible with viability), this means that life could go extinct in a hundred times shorter period—just in 10 million years.

This did not happen. Besides the program of biotic regulation, the genetic information of life also comprises a program preventing its own decay (erosion). The orderliness of living systems is maintained by a mechanism that has no counterparts in the inanimate world. This mechanism is among the key features that differentiate life from non-life.

All living objects form populations. Individuals of a given biological species are all similar to each other, which is why they can be assigned to a particular species. But there are no species composed of just one individual! Within each population individuals compete with each other. This competitive interaction reveals individuals with eroded genetic programs leading to deviation from the species' behavioral and morphological norm. Such individuals are forced out from the population in one way or another, while copies of individuals with normal genetic program fill the vacancies. It is this mechanism that prevents the loss of order in living objects at different levels of organization, from cells to local ecological communities (Gorshkov and Makar'eva 2001).

Many animals form internally correlated social structures, where individuals share a communal living (for example, bee hives, ant hills or tribes in mammals). Within a social structure all individuals continuously interact with each other. In the course of competitive interaction the social status of all individuals is determined. Individuals with a lower than average competitive capacity get a low social rank but remain within the social structure and are not eliminated unless their competitive capacity drops below a certain threshold. Such hierarchic social structures represent a peculiar form of correlation between individuals that can be compared to correlation of cells within a multicellular body. Information about the internal correlation of the social structure is contained in the genetic program of the species. It is maintained by competitive interaction between different social structures, with defective structures eliminated from the population of such structures. Noteworthy, in theoretical biology related concepts raise heated debates (e.g., Nowak et al. 2010; Nowak et al. 2011; Gintis 2012), with the alternative (conventional) view being that internal correlation of living systems (like bees in a beehive) could be explained by selection acting on individuals. While a detailed discussion of these problems is outside the scope of this paper, we note that a crucial point missed in the conventional evolutionary paradigm is the need to explain genetic *stability* of such internally correlated structures rather their appearance in the course of evolution.)

We conclude that decentralized organization is key for order maintenance in life. Decentralization presumes lack of correlation in the functioning of living objects. In a centralized system like for example a beehive or a multicellular organism, the various parts of the system strongly depend on each other (they are internally correlated). One part cannot win when another part loses (e.g. brain in a multicellular organism does not benefit from a kidney failure). In a decentralized population of independently functioning living objects the situation is different: if a certain object loses functionality this does not impair functioning of the neighboring objects. For example, if a certain object had a poor program of coordinated behavior (e.g. if the foraging bees cease to feed the queen or if cells in a multicellular body start proliferate on their own forming a cancer tumor), such an object perishes, and its place in the ecosystem is occupied by the progeny of other, normal, objects.

Once an object becomes globally correlated such that competitive interaction becomes impossible, such an object is prone to disintegration and decay, whatever the nature of the object is. Thus, life cannot exist in the form of a single globally correlated super-organism Gaia, as Lovelock called it (Lovelock 1988). This is a consequence of the unique complexity of living objects: their orderliness cannot be maintained merely by interaction with the inanimate world which is virtually disordered compared to life. Rather, order maintenance is an intrinsic property of life itself.

Regulation of global environmental conditions, e.g. atmospheric CO_2 concentration, by the biota does not require a global correlation of life. The highest level of organization of life—local ecological community—is the elementary operational unit of biotic regulation (Gorshkov 1995; Gorshkov et al. 2000, 2004). In forest ecosystem local ecological community is represented by a tree or a group of neighboring trees together with the accompanying local microbiota: bacteria, fungi, small animals, which function in a coordinated manner similar to cells within a multicellular organism. For example, the understorey herbs with help of the network of mycorrhizal fungi connecting individual plants within the local ecological community can share their stores of carbohydrates with the tree to facilitate tree awakening from the winter season (Lerat et al. 2002, see also Van der Hejden and Horton 2009).

Every local community tends to stabilize its local environment towards the optimum. For example, if CO_2 concentration is too high, the community will act to remove CO_2 from the atmosphere and deposit in chemically inert compounds such as, for example, live biomass. Because of global mixing, a local community is not able to fully control local CO_2 concentration. However, if the small relative change of the local environment is sufficient to impart competitive advantage to the community, then all local communities will perform the same stabilizing function. This will result in a global inflow of CO_2 from the atmosphere into the refractory reservoir (organic or inorganic carbon). The sensitivity of the ecological community to small relative changes of local environment is a fundamental parameter of biotic regulation (Gorshkov et al. 2000).

3.3.6 *Information Losses During Evolution of Large Animals*

We now turn to the role played by large animals in the life-environment interaction. As we shall see, the principles of life-environment sustainability can provide useful information relevant to the problems of modern civilization.

Animals interact with their environment mostly via cells at their body surface. With increasing body mass the relative number of such surface cells diminishes inversely proportionally to the linear size of the animal: $(S/s)/(V/v) = l/L$, where S/s is the number of cells with surface area $s = l^2$ on the body surface of area $S = L^2$. Here l and L are, respectively, linear sizes of an average cell and the animal, V/v is the total number of cells of cellular volume $v = l^3$ in animal body of $V = L^3$.

In particular, for large animals with $l \sim 50$ μm and $L \sim 0.5$ m, the share of surface cells is just one ten thousandth (10^{-4}). Information flux in the animal body is proportional to total energy consumption of the animal. The larger the animal, the smaller is the share of its energetic and information flux it can spend to participate in the biotic regulation of the external environment, on which the animal depends. Large animals use the available fluxes of energy and information almost exclusively to maintain the orderliness of their internal milieu rather than external environment.

Living matter is characterized by a universal rate of energy consumption per unit volume (Makarieva et al. 2008). Thus across evolutionary domains the total energy consumption of organisms grows proportionally to the cube of the linear body size, while energy consumption per unit area of the ground surface occupied by the organism grows directly proportionally to the linear body size. Per unit area of the ground surface large animals consume an energy flux that is several orders of magnitude higher than the solar energy flux consumed by life. For example, a human body with a metabolic power of 150 W and area of the body projection of about 0.5 m^2 consumes about 300 W/m^2, which is 3000 times larger than the global mean power of photosynthesis 0.1 W/m^2 (Note that in trees only the surface cells of leaves, roots and cambium are active, while the bulk of wood is, unlike animal bodies, biologically inert and does not consume either energy or information. That is why the effective linear size of the metabolically active parts of the trees is very small.).

To summarize, an increase in animal body size leads, first, to a higher dependence of the animal on the environment owing to rising energy consumption per unit body surface area. Second, it leads to a decrease of the share of consumed energy flux that the animal spends on the regulation of the external environment. In other words, during evolution of large animals the genetic information about their interaction with the environment was continuously being lost. This process can be compared to reduction of various organs in parasitic species—for example, some parasitic worms lost their digestive system (and the corresponding genetic information from their genome) exploiting instead the internal milieu of their hosts. By analogy, large animals can be viewed as parasitizing on the environment maintained for them by the rest of the species of the biosphere.

3.3.7 Why Do Large Animals Exist?

Having lost a major part of the original information about environmental regulation large animals nevertheless enjoy a nearly ubiquitous presence in the biosphere. This suggests that the regulating part of the biota for some reason keep them in existence, and that they do play a certain role in biotic regulation. Surprisingly, this role is related to the ability of large animals to destroy biomass, at the expense of which they exist.

Physical destruction of biomass of the regulatory part of the biota is a rare event. It may happen in the result of physical catastrophes like volcano eruptions, hurricanes, tornadoes, windfalls, and fires. The regulatory part of the biota has a program of self-recovery after such physical disturbances. This recovery is performed by professional "species-repairers" whose population densities under normal undisturbed conditions are low. Such species can be compared to populations of T-cells in our blood. In the boreal zone conifers predominantly belong to the regulatory part of the biota, while species-repairers are represented by deciduous trees like birch, aspen, alder and various herbs and shrubs. After disturbances these species restore the environment to a state optimal for the regulatory biota. They thus work to their own disadvantage, as they change the environment in a direction that eventually is unfavorable for them. For this reason, when the optimal environment is restored, population densities of species-repairers radically decline. But these species must not disappear altogether. Otherwise there would be nobody to restore the environment after infrequent but catastrophic physical disturbances.

Physical disturbances arise infrequently and unpredictably. Long periods of time can pass without such disturbances affecting a given region. During such periods population densities of species-repairers could drop below a certain critical threshold, when the intensity of competitive interaction weakens and the genetic information of the species deteriorates. Large animals help prevent such a scenario as they destroy the regulatory part of the biota in a more regular way independent of physical processes. Introducing disturbances to the vegetation cover large animals create favorable conditions for the existence of plants-repairers. Such plants increase their population densities in areas of such disturbances (e.g., animal-made lawns, paths etc.).

Since humans are also large animals, the genetic program of our species must also carry information about how to destroy the regulatory part of the biota. As a manifestation of this program, humans thrive in recently disturbed areas inhabited by plants-repairers which humans, as well as other large herbivorous animals, use for food. These are relatively open green landscapes along riverbanks, lakes and seashores that closely border with the undisturbed regulatory biota (the so-called climax or primary forest). Views of such landscapes, often shown in famous paintings, bring about positive emotions in the majority of people (Haber 2004). Few famous paintings show undisturbed climax forest (see for example paintings by Ivan Shishkin)—it is apparently not the optimal environment for *Homo sapiens*.

3.3.8 Ecological Rights of Animals and Man

For biotic regulation of the environment to be stable, an important condition is that population densities of such species-destroyers (large animals) do not rise above a certain safety threshold. The regulatory part of the biota (trees, bacteria, fungi and small animals composing local ecological communities) is organized in such a manner that the share of energy consumption available to large animals is strictly limited. In stable ecosystems the share of ecosystem primary productivity consumed by all larger animals combined (from mice to elephants) does not exceed 1 % and rapidly declines with increasing body size of the animal (Fig. 3.3). Modern humans have exceeded this cumulative threshold by about an order of magnitude: with an account of wood consumption, cattle fodder and food for people, our civilization consumes about 10 % of global net primary productivity of the biosphere.

Since high population densities of large animals are not compatible with ecosystem stability, the ecological restrictions for large animals to keep a low density must be genetically encoded. This limitation takes the form of dependence between the animal home range and body size (Fig. 3.4). Animal home range, an individual territory where no aliens are generally tolerated, is approximately proportional to animal body mass. The larger the animal, the larger territory it must possess to normally function. Territory deprivation results in physiological disorders in many species, from tiny jerboas to rhinoceros. For instance, in captive rhinoceros populations those animals that were kept in open areas (where they could just see, albeit not move across, a large territory) reproduced better than those enclosed by high walls (Carlstead et al. 1999). This means that the territorial requirements of the species are genetically encoded on the physiological level: a visual signal that a large territory is potentially available is essential for facilitating reproduction process in the species.

As one can see from Fig. 3.4, the ecological right of humans to have a large individual territory of about 4 km^2 is significantly violated in modern densely populated societies. As with other large animals, there are all grounds to expect that overpopulation has had profound impact on human physiology and behavior.

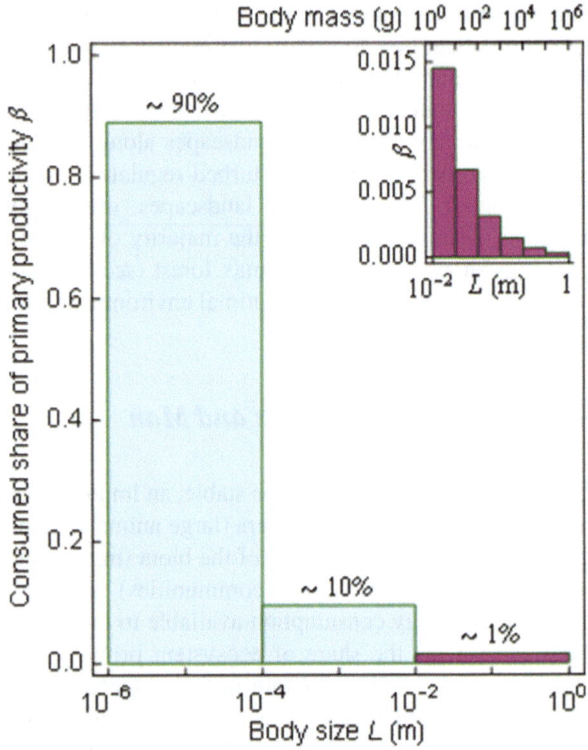

Fig. 3.3 Distribution of consumption of plant production in stable ecosystems. Unicellular organisms have controlled energy consumption at all times from the very beginning of life: in the modern biosphere over 90 % of plant production is consumed by the smallest organisms (bacteria and fungi). Arthropods, the smallest mobile animals, consume about 10 % of primary productivity. Ecological function of insects is similar to that of immune system: invasions of locusts, bark beetles etc. destroy defective plant communities. Insects are also important as pollinators. Dark pink diagram shows consumption of forest herbivores (mammals and birds) in the boreal zone (Makarieva et al. 2004b)

Data on human evolution appear to support this view (Knauft 1991; Boehm 2000). For the most part of our history as a species, humans existed in small groups containing about 30 individuals. Unlike great apes and more recent human societies, those simple human groups represented egalitarian and decentralized societies where all adult males were equal. Studies from such simple societies preserved until recently in the least disturbed tropical areas (e.g. in Papua New Guinea) confirm that the moral code in such groups discouraged dominance and hierarchy. These egalitarian societies, which represent a puzzle for anthropologists and evolutionary biologists, were characterized by low population densities and practically absent inter-group aggression. In other words, wars and massive killings of conspecifics were absent. In more recent societies where population densities rose to 2–3 person/

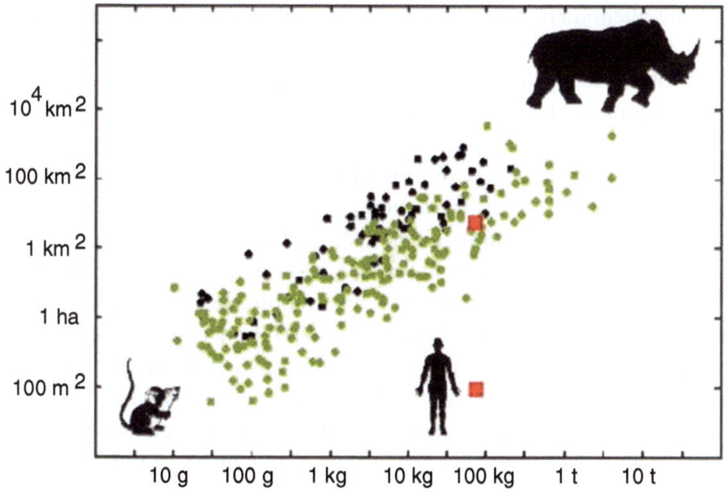

Fig. 3.4 Individual territories of mammals in their natural environments (*green symbols*—herbivores, *black symbols*—carnivores) as dependent on body mass (Data of Kelt and Van Vuren 2001). Humans were endowed by nature with at least 4 km^2 per capita

km^2 (i.e. where the ecological threshold of 0.25 ind/km^2 was exceeded by about one order of magnitude, Fig. 3.4) the egalitarian social order was replaced by a strict hierarchical structure with male-over-female dominance and individual property accumulation. It was in these overpopulated conditions that inter-group aggression (wars), people by people subordination and enslaving became part of human life which they unfortunately remain today. We will now discuss which human features might be held responsible for this development.

3.3.9 Principle Differences Between Animals and Man

Cellular processes in all living organisms proceed according to the genetic program encoded in DNA macromolecules. For plants, bacteria and fungi this is the only source of information to govern organism functioning. Plants, bacteria and fungi lack head and brain.

Locomotive animals cannot live on the basis of their genetic DNA program alone. Locomotion necessitates acquisition of additional information about new places visited earlier by the animal. This information accumulates in brain and is stored as memory. Animal behavior is governed by the genetically encoded positive and negative emotions that ensure animal's life in its natural environment. The animal tends to perform actions associated with positive emotions (sex, feeding) and escape actions entailing negative emotions (angst, pain). However, environmental factors that bring about these genetically encoded emotions can vary during the animal lifespan and be different for different generations in a population. This underlies the phenomenon of volition: the animal can undertake immediate actions associated with negative emotions ("to overcome itself") in order to experience positive emotions later. Information about factors that can bring about such delayed positive emotions is stored in the animal memory. This effect is used in animal training.

In animal species the genetic information of the DNA is transmitted to the next generation, while the information of memory accumulated during animal lifespan is lost with the death of the animal. This one-generation cycle of information accumulation and erosion contributes to the stable existence of species in their environment. The genetic information of DNA macromolecules has been tested on many generations of animals: it guides how the animal interacts with its environment in a sustainable manner and can remain practically constant during the entire period of species existence of the order of several million years. Meanwhile information accumulated in the memory inevitably contains some false elements that can prove useless or detrimental to the next generations. Thus memory must vanish with the death of the animal.

Homo sapiens is the only animal species who violates this rule. Memory information that accumulates during life of one individual is shared with and is assimilated by the next generations. This additional information of memory with trans-generational transmittance comprises the human culture. Indeed, humans are different from animals in having a culture (e.g., Kesebir et al. 2010). Cultural information, like individual memory, contains false and detrimental elements, but also useful elements enhancing population stability at least on a certain time scale. Some of these useful elements take the form of mystic and religious rules and social dogmas governing people's behavior.

The variable cultural information is inherently in conflict with the invariable genetic information that determines the strategy of the human behavior. False and detrimental cultural information could often make whole populations perish. Those cultural elements propagated contributed to or were compatible with social stability over a larger number of generations. People used cultural knowledge to survive on new territories by transforming local biota to a state resembling their natural ecosystem, for example, when replacing natural forests by pastures and agricultural fields. Cultural knowledge was used to exploit the biosphere more and more intensely. As the natural biota degraded and global environment started to lose its stability under the growing anthropogenic impact, those cultural rules that were used to stabilize the society in a globally stable environment, became destructive and threatening the existence of the civilization in a new, changed environment.

In the modern world specific social structures propagated in territories now termed countries. Territorial integrity of countries whose population comprises different ethnic groups and nationalities is maintained by what can be termed the culture of multi-ethnic patriotism. However, cultural integrity in a large population is unstable and spontaneously disintegrates into the genetically programmed cultural integrity of small social structures containing the normal (low) number of members. Therefore, the culture of patriotism demands continuous efforts on its maintenance and propagation in the younger generation.

3.3.10 Scientific and Technological Progress and Ecological Human Needs

In animal populations in the course of competitive interaction the least competitive individuals are forced outside the species range where they cannot live normally (they either die or survive but do not leave progeny). Individuals of *Homo sapiens* forced out of the natural environment where our species came to existence did not perish but were able to spread all over the world owing to the accumulated cultural knowledge.

However, human existence outside the natural species-specific environment is associated with excessive physical exercise and emotional stress. When humans colonized high latitudes with their low biological productivity and unfavorable temperature regime they had to exert more physical efforts to obtain food and maintain optimal temperature in their homes. Ecosystem productivity in the tropical zone is about three-four times higher than the productivity in the temperate and boreal zones where the modern technological progress was born. This gives an idea of the very significant amount of overworking that *Homo sapiens* individuals had to experience outside their natural ecosystem. This genetically encoded dissatisfaction with the unfavorable environment determined the direction of the technological progress, of which there was no need in human populations remaining in their optimal environments in the tropics.

From this perspective, one cannot expect the technological progress to be improving human conditions for ever. If the technological progress has a start (the moment when a sufficient number of humans were forced out of their natural ecosystem) and a cause (these humans were genetically dissatisfied with their new conditions where they had to work too much), it must also have an obvious end—when the human needs that motivated this progress get satisfied and the motivation disappears.

If we look at the major achievements of the technological progress in the last two centuries we notice that to a large degree they were aimed at freeing people from hard physical labor. In the pre-industrial era a major difference in lifestyle between the poor and the rich was that the poor had to perform hard work, while the rich did not. From the viewpoint of the genetic program of our species this was a fundamental difference: some individuals had to overwork exhausting their biophysical capacity, while the others did not. In the industrial era this fundamental difference was practically erased by a wide variety of technical aids.

Today almost all people in the developed world have running water, central heating, various electric appliances to facilitate housekeeping and to permit cars to get moved around. In the result, the vast majority of the population has been deprived of the ancient stimulus to participate in the technological progress. Unsurprisingly, the remaining islands of rapid economic growth today are concentrated in those regions where a major part of the population is still engaged in rough physical work (China, India, African countries). Another indication that life style improvements have reached a plateau is the dynamics of leisure time change in the developed countries. Aguiar and Hurst (2006) showed that in the United States in 1965–2003 the leisure time increased most in those population cohorts that were primarily occupied with hard physical labor (low educated males).

Indeed, today many professionals voice skepticism concerning the future potential of the technological progress to improve human life (Cowen 2011; Atkinson and Ezell 2012; Mokyr 2013; Gordon 2014). However, few existing analyses view the apparent technological slowdown in a broader scientific perspective that would include ecological human needs. For example, the middle income trap, a major obstacle for economic growth perceived by economists (Kharas and Kohli 2011), appears to be readily explainable in terms of a greatly reduced motivation for working efforts after a majority in the population have freed themselves from hard labor.

While the idea of a hierarchy of human needs has been highly influential in modern thinking (Peterson and Park 2010), the ecological human needs associated with overworking and lack of appropriate individual territory have been invariably ignored. Physiological needs residing at the bottom of the pyramid have been traditionally considered as being the most "straightforward", almost invariably exemplified just by hunger and thirst (Kenrick et al. 2010; Ackerman and Bargh 2010; Lyubomirsky and Boehm 2010). Meanwhile in the more comprehensive picture that we have presented the last two centuries appear as a point of singularity in human history, because it is for the first time that technological progress eliminated the need for exhaustive physical labor in the majority of human population. One of the essential ecological rights of *Homo sapiens* was for the first time in human history satisfied outside the natural ecological niche.

However, at the same time the other equally essential ecological rights of humans remained unsatisfied or became violated. In particular, owing to the exponential growth of population density and total number of members in social structures people lost their rights for an appropriate individual territory and social significance. A major human right that was respected in the natural environment is the right of any large animal to move, using muscle power, over an individual territory free from aliens and competitors (Personal car transportation is popular with modern humans as it creates an illusion of the possibility of such movements). This right was violated in our species at an unprecedented, global scale. Individual territories of modern people are comparable to individual territories of shrews (Fig. 3.4).

Continuing automation turns labor, which in the right amounts is necessary for the normal human existence, to a privilege and at the same time into a deficit. As a growing number of people become unemployed, they lose the right to participate in the maintenance of their society and thus lose their social significance. The deficit of social significance aggravates. The explosive development of Internet, mobile connections and social networks, which were largely responsible for the global economic growth of the last two decades, made profit exactly from this deficit of social significance. With help of the Internet it became possible for people to group by interests and form small social structures (reference groups) with their size resembling that of the normal social structure of humans.

There is principle difference between scientific and technological progress. Scientific progress reflects the accumulation of objective knowledge about the external world. It is based on the genetically encoded ability of humans to accumulate culture. People experience positive emotions when they get to know something new about the world irrespective of the field of knowledge to which the new information belongs. Scientific research is continuously generating an enormous amount of new knowledge. From this treasure of knowledge technological progress selects information that could be used to satisfy human needs and make the artificial environment resemble natural environment of our species. Great scientists and engineers who made outstanding discoveries opening new horizons for technological progress appear in the human population very infrequently. Fundamental breakthroughs like the discovery of electricity or invention of the Internet did not lead to financial prosperity of the creators. Rather, the new inventions are brought to mass culture by active entrepreneurs. In the large global population of individuals unsatisfied with their living conditions the number of such entrepreneurs has always greatly exceeded the number of new potentially useful inventions.

Nowadays there appear few remaining ways in which technological progress could satisfy real human needs: its potential has been almost exhausted. Practically, technology has been able to improve human lives in but one essential way—it freed people from rough tiring labor. In this situation it is quite useless to call for an increase in the buying capacity of the consumers and consider them as the main drivers of technological progress and economic growth (Hanauer 2012). As we discussed above, what modern consumers might wish to buy to live a satisfactory life, worthy to human beings the technological progress can hardly offer anymore.

The only direction of modern technology that remains of real interest to mass consumers is medicine, which appeals to the fundamentally insatiable genetically encoded human instinct of self-preservation. It is for the first time in human history that technological progress caters mostly for the needs of the sick and the elderly who continue to play a significant role in the society.

In modern world per capita energy consumption is about 2.5×10^3 W, which exceeds the biological energy consumption of a human adult, 150 W, by more than one order of magnitude. One can say that every person has more than ten servants—robots working with a power equal to the power of an adult man. With increasing automation of all the spheres of life (term the Internet of Things), increasing pension age and decreasing load by children, the increase of the mean population age does not pose any economic problems. (Indeed, even in primitive societies people are able to support themselves up to the age of 60–65 years (Kaplan et al. 2000).) Only those concerned about the decelerating economic growth perceive this increase as catastrophic. However, as we discussed above, global economic growth very likely will cease in any case.

At the same time as the global resources become depleted this global challenge creates a novel stimulus for technological progress. If the global stability of the environment is not lost, technological progress will be directed at maintaining modern living standards in the situation of aggravating shortages of all resources and degrading global environment. This global role of technology is new and distinct from what it used to be until recently—improving (at least some aspects of) human conditions rather than merely sustaining the status quo.

3.3.11 Conclusions

Considerations of the major principles behind sustainability of living systems allow us to formulate a strategic vision on the global problems of the humanity as well as on their possible solutions. First of all, our considerations urge a significant shift in the direction of globally centralized efforts aimed at preserving the global environment and climate. We have discussed how natural ecosystems perform regulation of the global and regional environment and climate, which is a self-sustainable, ultra-complex and highly energy-consuming planetary process that cannot in principle be replicated by technology (Sects. 3.3.2 and 3.3.3). Thus, to preserve a climate suitable for life on Earth we need to preserve and restore natural ecosystems in the first place, allowing them to perform their work for the benefit of the humanity and life as a whole.

The prevailing view on environmental problems in the modern society appears to have a different focus. Environmental issues are understood by modern society as the challenge to protect the environment from technological pollution. One assumes that if the technological cycle becomes closed based on environmentally friendly and renewable 'green' energy, such that the pollutants including CO_2 emissions are quantitatively diminished, the environmental problems will be solved. Such a view underscores an entirely different strategy of coping with the global crisis: for example, it can encourage elimination, rather than preservation, of natural forests in favor of growing biofuel. Within this perspective, the transition to renewable energy sources and recycling is thought to be able, at least in principle—neglecting the practical limitations (e.g., Abbasi and Abbasi 2012), to overcome environmental problems and thus lift any limitations on further growth of global economy and population.

However, if we take into account that the environment on Earth is under biotic control, recycling and renewables are not a strategic option but may in some cases significantly aggravate the environmental situation. The only possibility to preserve an environments suitable for humans is to reduce the anthropogenic consumption of the biosphere resources down to the natural threshold of about one tenth of per cent to preserve biotic regulation on a global scale. This means that the modern rate of the anthropogenic consumption of primary productivity and consequently the global population number should decrease by two orders of magnitude. Energetic needs of such a population, where the right for appropriate individual territory will be respected (see Fig. 3.4), could be fully met by the hydropower—theoretically at least—which is the only renewable energy flow that can be exploited by people without a continent-scale destruction of the biota. The per capita energy consumption can remain the same or even grow.

Recognition of global overpopulation and degradation of the remaining natural ecosystems as a major threat to global and regional conditions will demand a centralized program of responding to this challenge. How can we ensure that such a program, if it is proposed and agreed upon, makes practical sense and does not lead to dangerous outcomes?

At this point we come back to the issue of the right balance between central and decentral solutions. As we discussed in Sect. 3.3.4, "centralization" is equivalent to "internal correlation" of an object, while "decentralization" is equivalent to local independence of the various objects or parts of the object. In the context of human civilization, an object might be equivalent to the population of a country, a region, a city or a village, whereas the parts of the object are constituted by individual people, the population of a region, a city or a village. Moreover, to a large degree our civilization itself represents an internally correlated object that, unlike any other object in living nature, exists in a single number. As such, it is a priori highly vulnerable.

Decentral solutions are based on the knowledge of and respect for climatic, economic and social conditions and competences at the local scale. If some local strategy fails, the country (or region, or civilization as a whole) will not collapse. The local community of people will suffer but they will be able to borrow more efficient local solutions from their more successful neighbors. In this way good solutions will spread, while bad solutions will be abandoned without threatening the society (life) as a whole. Therefore, it is not just desirable but vital to delegate all functions that can be locally maintained to local communities. The subsidiarity principle (Vause 1995; see Sect. 2.2) takes care of such local interest and is therefore associated with decentralization. It suggests that the central authority should support, rather than subordinate, local functions. It should perform only those tasks which cannot be performed effectively at the local level.

The situation of finding global, centralized solutions presents a significantly greater challenge. For a single object like global civilization there is no opportunity to test in an experiment, using some other objects, whether the strategy that we as a planetary community choose is salvaging or suicidal. In natural ecosystems new meaningful genetic information that ensures centralized functioning of a system to the benefit of all of its parts does not emerge all of a sudden in the face of some environmental challenge. Such new information appears in the course of the evolution by an infrequent chance—in the result of mutations the overwhelming majority of which are harmful, but very few turn out to be useful. Once established, the information about a successful centralized strategy is further prevented from disintegration by competition of many centralized but mutually independent structures.

During evolution of large animals, of which *Homo sapiens* is an example, the genetic information about environmental regulation got lost (Sect. 3.3.5). Moreover, the genetic program of *Homo sapiens* as well as of other large animals prescribes a behavior that to a certain degree destroys the biotic environment (Sect. 3.3.6). If the humanity possessed unlimited energy sources then following this genetic program would have led to a complete degradation of natural ecosystems. As a consequence, the environment that is favorable for humans would have been also destroyed on a global scale. Degradation of the biotic pump on deforested land would have led to disruption of the water cycle, while overexploitation on land would have resulted in irrecoverable soil erosion. Therefore, we conclude that the program for centralized actions to avoid a global collapse cannot be formulated solely on the basis of intrinsic human desires, instincts and emotions that are all governed by the genetic program of *Homo sapiens*.

However, humans are unique organisms whose life is governed not only by genetic but also by cultural information (Sect. 3.3.8). Cultural information of the humanity comprises, along with other elements, objective scientific information which is not subject to pluralism. Scientific information is truth, because it is checked for its concordance with observations and scientific trials. So there is only one opportunity to find a successful centralized strategy for the civilization as a whole: to derive it from the best available scientific knowledge synthesized across disciplines into a coherent, non-contradictory framework. This is a new challenge for the intellectual elite of the Earth.

Until recently the achievements of science were judged by their ability to enhance human transformation of the biosphere that was accompanied by a rapid population growth, destruction of the natural biota and its regulatory environmental potential. Thus, the humanity has got into the present critical situation, which threatens the very existence of the civilization, owing to science and technology (Sect. 3.3.9). However, now it is only with help of comprehensive scientific knowledge about the humanity, that our planet and life as a whole get a chance to overcome the current global environmental crisis and preserve our civilization.

3.4 Nature and Human Nature: Ethical Concerns Should Not Be Disregarded in the Process of Decentralization

Verena Risse

3.4.1 Key Messages

Global problems often have local effects and demand local action. This supports the claim that it could be advisable to envisage decentralized solutions. Still, ethical concerns may not be ignored when this process is implemented.

3.4.2 Introduction

Among the most pressing problems today are the environmental degradation and political instabilities. Both problems are global insofar their causes and effects have a planetary dimension that knows no borders. At the same time, the effects become visible in concrete places and situations. Therefore, considering the principle of decentralization to approach these problems seems reasonable.

When it comes to finding solutions for global problems that do not concern humans alone, but also their environment, several interventions at the wbk3 suggested to investigate nature for inspiration. Indeed, nature as a complex system that has survived, evolved and got adapted over billions of years shows a variety of patterns of how to deal with changes and obstacles.

Human beings are part of nature and share many characteristics with other living species. Yet, they are also special in that they are—at least as far as we know—the only species granted with moral insight and the capacity for ethical reflection. This moral capacity is not merely something created to occupy philosophers. Rather it is a capacity that is crucial for the survival of mankind, especially for people living together in communities.

This contribution therefore argues that ethical considerations may not be ignored whenever new technical, social or political solutions are developed—even if this is done with an inspiration from nature itself. To make this point, the first section recalls some of the global problems and considers why decentralized solutions might be pertinent. The second section lays out why ethical concerns must be taken into account and what this can amount to in different situations.

3.4.3 Global Problems: Local Effects

The world faces a considerable number of problems most of which are man-made. Some of these problems like environmental degradation, climate change or diseases like Ebola significantly determine the lives of humans today and of future generations as well as they deeply affect the animal and plant life. Moreover, there are signs that in some cases, nature has lost its capacity to recover from or to adapt to the changes it suffered. A case in point is the increasing mortality of certain submarine species in the warming Mediterranean (Rivetti et al. 2014). Also in the social and political dimension, a new degree seems to be reached. Just think of spreading terrorist networks, more refugees than ever since the Second World War (UNHCR 2014) or rising social inequalities between and within countries.

Both sets of problems, the environmental and the socio-political ones, are global in scope. This means that they are affecting the entire planet and are not bound by the territorial borders of a state. Take the case of climate change. CO_2 emissions are considered the major cause for climate change and CO_2 emissions cannot be held within the territory of those who produce them. This example also suggests that the problems are not only global, but that the line between mere natural and socio-political problems is hard to draw. As in the case of climate change, often it is not the countries that emit most CO_2 that suffer most from its effects. This therefore prompts an issue of inequality and by extension of injustice that deserves attention (Caney 2005).

Moreover, while these problems are global in scope, their effects are experienced locally. This is to say, due to rising sea levels specific islands are flooded and the habitat is being destroyed. Likewise the refugees arrive at and seek asylum in a specific state. Therefore, it seems pertinent to address these problems in a decentralized fashion. What this can (technically) amount to is the subject of other contributions to this volume and will therefore not be explored here. Instead, the following section intends to show why ethical reflections should accompany the process of finding and implementing (decentralized) solutions.

3.4.4 Taking into Account Ethical Concerns

We can not only find inspiration from nature as to how to improve technical or institutional solutions, we can also learn something about ourselves as humans. This is to say, insights from evolution theory remind us that human beings are unique in possessing rationality which involves the capacity to raise ethical concerns (already in Aristotle, (1998) *Met.* 103b1-2, 1041a25-32). In fact, this edited volume is just an illustration of ethical concerns being articulated. At the same time evolution theory suggests that acting according to these concerns is contrary to our nature, for we intuitively choose the easier, cheaper, faster way which is not always in accordance with what moral behavior demands. This implies that behaving morally often means making an effort and perhaps overcoming one's inclinations.

These difficulties notwithstanding, it must be born in mind that the ethical concerns arise for a reason. This means that solving the global problems does not merely depend on finding the before-mentioned solutions—be they centralized or decentralized—even if these are inspired by nature itself. Instead, ethical considerations are likewise important and pertinent in several respects.

First, they are necessary to clarify the relations between human beings as well as their relation with nature and other species. This may include questions such as who is responsible for certain degradations, how are the costs and benefits of technical solutions to be distributed, should some areas or resources be preserved or granted to indigenous groups or future generations etc.

Besides this socio-environmental dimension, ethical considerations ought not to be ignored, secondly, because they can help individuals to get a grip on their own role, concerns and rights within the process of adaptation. This, for instance, involves how to deal with the loss of a job, a home, maybe even a whole home country as happens to inhabitants of certain island states due to rising sea levels.

Finally, it must be stressed that these ethical questions are not to be regarded as a separate endeavor discussed by some specialists only. Rather, they are present in all aspects of life including technology and should therefore be treated there, too. Neither are ethical concerns bound to a specific institutional structure or level, so that they can follow the request for decentralization. Quite the contrary, it is important to bear in mind that the opposition between local and global can rarely be upheld as many seemingly 'global' decisions are locally rooted. The most obvious example is perhaps that UN staff as global decision-makers work in offices which are located in New York City (Scholte 2008). In this sense, also ethical reflections such as the ones articulated here have their place and location.

3.4.5 Conclusion

This contribution has not argued for a particular set of ethical principles. Rather, the argument is located at an earlier stage in that it stressed the importance to apply ethical considerations and to make them a part of those processes of adaptation or decentralization that are outlined in this volume.

3.5 Transformation Towards a Resilient and Humane Environment and Culture: What Needs to Be Done?

Carolin Böker, Bettina Haas, and Ortwin Renn

3.5.1 Key Messages

We need a transformation towards a development that emphasizes resource efficiency, effective governance structures, fair distribution of opportunities and resources, and a resilient approach to risk taking. This includes a special attention to underprivileged individuals, groups and nations and we need a global agreement on the goals that we want to accomplish with respect to a sustainable path to future development.

3.5.2 Introduction

The perception of humane living and working conditions differs between variable cultures and parts of the world. Access to sufficient food and safe drinking water, electricity, education, humane labor and medical care has been accomplished in most North American and European countries, whereas in many African, Asian,

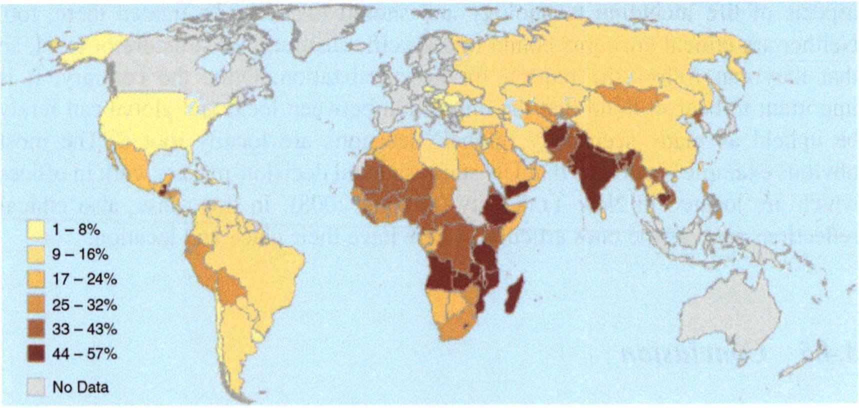

Fig. 3.5 Map of the world, different colours visualize the percentage of children under the age of five suffering from stunting (The Global Education Project)

Indopacific and South American countries these basic services are still lacking for a large proportion of the population. Taking Brazil as an example of South American countries, many children have to face maltreatment, street work, urban violence and abuse in Saõ Paulo, which correlates with poverty, single motherhood and a troubled family background (Mello et al. 2014). Additionally, due to these circumstances, children do not regularly attend school and thus often stay in bad living and labor conditions. Bad living conditions all over the world are also reflected by a high percentage of young children suffering from stunting (Fig. 3.5). In developed countries poor living conditions are more a concern to social underprivileged, often older people. In Germany for example, poverty among the elderly is not an exception: especially women are affected by the privatization of pensions with a reduction of the pension level at the same time (Fachinger 2008).

Poor working conditions might be divided into two different types of effects: physiological and mental maladies. Physiological effects include threats to health due to insecure work-flows, such as handling toxic substances without precautions and suitable safety equipment. A good example for negative physiological impacts on workers' health are the small-scale gold-mines in South American, Asian and African countries. Gibb and O'Leary (2014) reviewed the impacts of mercury due to the direct exposure during work (inhalation of vapours) and the indirect exposure after work (e.g. uptake of contaminated food such as fish). This can harm the nervous system, the intestine, the lungs and kidneys. In contrast to that, in developed countries, mental problems caused or promoted by a negative working environment (e.g. pressure of time and competition), are more and more evolving. Thus, selective serotonin-reuptake inhibitors as antidepressants are very popular in North America and Europe (Kirmayer 2002).

Taken together, all countries—undeveloped, developing, as well as developed ones—have to deal with problems concerning a humane environment for living and working together. We urgently need a transformation towards an economically just and socially fair development without disadvantages for underprivileged groups of people (e.g. children, elderly).

The International Expert Group on Earth System Preservation discussed some ideas how to get a step further towards humane living and working conditions all over the world.

3.5.3 How to Define "Humane Living Conditions"?

Pre-industrial societies lived more or less in a system of "subsistence"—i.e. the consumption and production of goods was mainly aimed at self-supply. Although trading took place in its pristine form, the main task was to ensure the living (subsistence), leaving little room for gaining or maximizing profit. Due to industrialization, the economic system changed to market economy or capitalism. Among many other characteristics, production and consumption are decoupled and maximizing profit is one main goal embedded in the economic system (Kilching 2008). Due to the digital revolution and political decisions (liberalization of trade), within the last 50 years the so called "globalization"—international exchange of goods and capital—even further decoupled production and consumption.

Societies living in a system of subsistence seem to be roughly in balance with available resources, as long as population growth was more or less constant. With the advent of capitalism and capital growth, the demand for ecosystem services and long-term resource availability have become more and more separated. Today, mankind uses statistically the resources of about 1.5 planets earth—going versus 2 planets in 2030 (Global Footprint Network 2014). It is obvious, that this cannot be in line with sustainability as well as humane living conditions.

The question "what are humane living conditions" is a very difficult issue and would demand a comprehensive answer that would reach far beyond the scope of this paper.

Nevertheless, some principal points seem to be beyond dispute:

- Drinking water and food in adequate quality and quantity
- Housing and sanitary facilities
- Health care
- Basic security (against violence, criminal acts, suppression, etc.)
- Access to education
- Work/job/occupation
- Contentment, happiness
- Self-determined living

These considerations are focused on the human individual or groups of individuals.

Another approach may be to postulate, that overexploitation of earth's resources in general is inhumane, as its logical consequence will be to render it very difficult (to impossible) for human societies to live on earth in distant future.

3.5.4 Basic Points of Transformation

The earth is home to many different terrestrial and aquatic ecosystems. Without anthropogenic impacts, these complex and sensitive networks would be self-regulating without any central supervision. Why not taking them as models for managing nations or regions was one of the questions of the third Meeting of the International Expert Group of Earth System Preservation. The answer was unexpectedly clear: ecosystems are suitable models to study networks and dependencies, but not for managing cities, provinces or even states (Haber 2010). Anthropogenic systems cannot be self-regulating, they need governance and guidance based on ethical values. How governance and ethics are organized and achieved depends on many different factors such as cultural backgrounds and geography. In short, there are two possible extremes to organize governance and abidance by the law: centralized and decentralized. What we need is a governance combining the advantages of both forms depending on the specific requirements that are time- and context dependent. On the one hand centralized systems, such as dictatorships, hold a certain amount of risk that the power is misused. On the other hand, centralized systems allow for quick and direct decisions, which can be an advantage in cases of pending catastrophes. In decentralized systems it is often vice versa: decision-making can be a time-consuming and a cumbersome bureaucratic process. However, a misuse of power is more unlikely or at least more difficult.

No matter of how centralized or decentralized a system is organized, the ultimate goal must be sustainability. With the increasing world population there is no sustainable future for humankind if we continue at the present consumption and exploitation rates.

The further depletion of ecosystems such as forests, oceans and streams must be averted. Additionally, these systems cannot be used as sinks for emissions and non-biodegradable waste (e.g. plastic, chemical, radioactive, toxic waste) anymore. Moreover, better recycling processes must lead to a reduced use of natural resources. The re-use of resources will probably be one of the greatest challenges to future generations.

The transformation to a sustainable management of resources must go along with the protection of individual rights accompanied by a governance system that includes self-determined living as well as the access to resources for basic needs (e.g. safe drinking water).

One way of how the transformation can be facilitated might be through better education programs including all age groups. The development of a broader understanding of environmental and social topics is necessary for preparing a mindset for sustainability and encouraging behavior that promotes sustainable practices rather than end-of-pipe solutions. In a social context sustainability means a pathway to a just and fair distribution of living opportunities and resources. This includes the need to minimize the gap between the very poor and the very rich.

3.5.5 Conclusion

Decentralized, self-regulating ecosystems may help to better understand anthropogenic networks, but ecosystems cannot serve as a normative model for humane behavior.

The fist condition for mitigating the environmental impacts of an unsustainable lifestyle (to surrounding nature as well as other humans) is human awareness and attention. Based on a general agreement that sustainability is the major goal of global governance, the second step is to create a governance structure that ensures resource efficiency, effective transformation steps towards sustainable living conditions, a fair distribution of opportunities and resources, and a resilient approach to risk taking. For that purpose the world needs externally established and controlled structures (good governance) in order to reach an improvement (Grambow 2013).

Significant changes in behaviour of the "world community" are necessary—and this within a comparatively short time frame. "Climate change" is one prominent example—the problem has been identified well in advance, but effective and consistent reactions are still lacking due to individual (industrial branches to whole states) egoism. Already now one can witness accelerating negative effects.

There are important principles that could guide humankind through the process of transformation, including:

- A balance between resources and environmental ecosystem functions
  - No depletion of oceans, forests, etc.
  - No overuse of the earth as sink for emissions and waste
- Respect for basic human needs
- Sufficiency as a goal for individual wealth accumulation
- Resilience with respect to risk taking
- Fairness with respect to opportunities and resources
- etc...

Today we do not face a lack of perception of the problem or ideas for potential solutions. For example *Nazrul Islam* 2013] (p. 2–3) states that a "new social model" is necessary, as ".... the current model is leading to breaches in planetary boundaries, jeopardizing the very existence of human civilization on this planet. ... the current model is not proving efficient for achieving human development goals in developing countries. ... the current model is not proving that helpful in improving life satisfaction in developed countries either" and describes the benefits of a change: "...acceptance of and steps toward the sustainable social model can bring environment and development together. The process has to begin with transition to sustainable consumption in developed countries. This transition will however require transformative changes in the economy, society, culture, and lifestyle. These changes will constitute a new phase of human development for developed countries. Thereby human development will become a universal goal applicable to both developed and developing countries ... The transition of developed countries toward sustainable consumption pattern will increase the resource and environmental space for developing countries to grow and improve their material standard of living. It will also have a demonstration effect by offering a different 'aspiration model,' so that developing countries may no longer strive to adopt the unsustainable consumption pattern currently observed in developed countries."

The most difficult and yet unanswered question is how to convince decision makers, economic and political leaders and high profile opinion leaders that the long-term sustainability of humankind depends on a radical transformation of established economic and social patterns towards a balance between demand and long-term availability of natural resources and a fair and equitable distribution of these resources among nations and individuals. This implies daring to re-think decision patterns, focusing on long-term effects, and not on short term economic gain.

The scientific community can offer support for this transformation by developing better interdisciplinary system sciences for exploring the impacts of human interventions into the natural and social environment. It can provide important transformative insights of how to pursue an effective path towards sustainability given the knowledge about systems and how they respond to interventions.

References

References of Section 3.2

Parsons TE, Shils EA (1951) Toward a general theory of action. Cambridge University Press, Cambridge

Renn O (2014) Das Risikoparadox. Warum wir uns vor dem Falschen fürchten. Fischer Taschenbuch, Frankfurt am Main, pp 498–523

Rosa E, Renn O, McCright A (2014) The risk society revisited. Risk governance and social theory. Temple University Press, Philadelphia, pp 173–178

References of Section 3.3

Abbasi T, Abbasi SA (2012) Is the use of renewable energy sources an answer to the problems of global warming and pollution? Crit Rev Environ Sci Technol 42:99–154. doi:10.1080/10643389.2010.498754

Ackerman JM, Bargh JA (2010) The purpose-driven life: commentary on Kenrick et al. (2010). Perspect Psychol Sci 5:323–326. doi:10.1177/1745691610369472
Aguiar M, Hurst E (2006) Measuring trends in leisure: the allocation of time over five decades. Working paper no. 12082, National Bureau of Economic Research. doi:10.3386/w12082
Anitori RP (ed) (2012) Extremophiles. Microbiology and biotechnology. Caister Academic Press, Norfolk
Aslak T, Bruaset AM (2013) The internet of things. In: Bruaset AM, Aslak T (eds) Conversations about challenges in computing. Springer, New York, pp 21–26. doi:10.1007/978-3-319-00209-5_3
Atkinson RD, Ezell SJ (2012) Innovation economics: the race for global advantage. Yale University Press, New Haven
Berner RA (1990) Atmospheric carbon dioxide levels over Phanerozoic time. Science 249:1382–1386
Boehm C (2000) Conflict and the evolution of social control. J Conscious Stud 7:79–101
Canganella F (2012) Ecology and biotechnology of extremophilic microorganisms, particularly anaerobic thermophiles. In: Singh OV (ed) Extremophiles: sustainable resources and biotechnological implications. Wiley, Hoboken. doi:10.1002/9781118394144.ch7
Carlstead K, Fraser J, Bennett C, Kleiman DG (1999) Black rhinoceros (*Diceros bicornis*) in U.S. zoos. II. Behavior, breeding success, and mortality in relation to housing facilities. Zoo Biol 18:35–52
Cowen T (2011) The great stagnation. Dutton, New York
Dean J, Ghemawat S (2008) MapReduce: simplified data processing on large clusters. Commun ACM 51:107–113. doi:10.1145/1327452.1327492
Dobzhansky T (1973) Nothing in biology makes sense except in the light of evolution. Am Biol Teach 35:125–129
Dombrowski U, Wagner T (2014) Mental strain as field of action in the 4th industrial revolution. Procedia CIRP 17:100–105. doi:10.1016/j.procir.2014.01.077
Downing K, Zvirinsky P (1999) The simulated evolution of biochemical guilds: reconciling Gaia theory and natural selection. Art Life 5:291–318
Free A, Barton NH (2007) Do evolution and ecology need the Gaia hypothesis? Trends Ecol Evol 22:611–619. doi:10.1016/j.tree.2007.07.007
Garrels RM, Lerman A (1981) Phanerozoic cycles of sedimentary carbon and sulfur. Proc Natl Acad Sci USA 78:4652–4656
Gintis H (2012) Clash of the titans. Bioscience 62:987–991. doi:10.1525/bio.2012.62.11.8
Gordon RJ (2014) The demise of U.S. economic growth: restatement, rebuttal, and reflections. Working paper no. 19895. National Bureau of Economic Research. doi:10.3386/w19895
Gorshkov VG (1995) Physical and biological bases of life stability. Man, biota, environment. Springer, Berlin
Gorshkov VG, Makar'eva AM (2001) On the possibility of physical self-organization of biological and ecological systems. Dokl Biol Sci 378:258–261
Gorshkov VG, Makarieva AM (2002) Greenhouse effect dependence on atmospheric concentrations of greenhouse substances and the nature of climate stability on earth. Atmos Chem Phys Discuss 2:289–337
Gorshkov VG, Gorshkov VV, Makarieva AM (2000) Biotic regulation of the environment: key issue of global change. Springer-Praxis, London
Gorshkov VG, Makarieva AM, Gorshkov VV (2004) Revising the fundamentals of ecological knowledge: the biota-environment interaction. Ecol Complex 1:17–36. doi:10.1016/j.ecocom.2003.09.002
Haber W (2004) Landscape ecology as a bridge from ecosystems to human ecology. Ecol Res 19:99–106. doi:10.1111/j.1440-1703.2003.00615.x
Hanauer N (2012) Rich people do not create jobs. Censored TED talk available at: www.youtube.com/watch?v=CKCvf8E7V1g

Kaplan H, Hill K, Lancaster J, Hurtado AM (2000) A theory of human life history evolution: diet, intelligence, and longevity. Evol Anthropol 9:156–185
Kenrick DT, Griskevicius V, Neuberg SL, Schaller M (2010) Renovating the pyramid of needs: contemporary extensions built upon ancient foundations. Perspect Psychol Sci 5:292–314. doi:10.1177/1745691610369469
Kesebir S, Graham J, Oishi S (2010) A theory of human needs should be human-centered, not animal-centered: commentary on Kenrick et al. (2010). Perspect Psychol Sci 5:315–319. doi:10.1177/1745691610369470
Kharas H, Kohli H (2011) What is the middle income trap, why do countries fall into it, and how can it be avoided? Glob J Emerg Mark Econ 3:281–289. doi:10.1177/097491011100300302
Knauft BM (1991) Violence and sociality in human evolution [and comments and replies]. Curr Anthropol 32:321–428
Lerat S, Gauci R, Catford JG, Vierheilig H, Piché Y, Lapointe L (2002) ^{14}C transfer between the spring ephemeral *Erythronium americanum*and sugar maple saplings via arbuscular mycorrhizal fungi in natural stands. Oecologia 132:181–187
Lovelock J (1988) The Ages of Gaia. Oxford University Press, Oxford
Lyubomirsky S, Boehm JK (2010) Human motives, happiness, and the puzzle of parenthood: commentary on Kenrick et al. (2010). Perspect Psychol Sci 5:327–334. doi:10.1177/1745691610369473
Makarieva AM, Gorshkov VG (2001) The greenhouse effect and the stability of the global mean surface temperature. Dokl Earth Sci 377:210–214
Makarieva AM, Gorshkov VG (2007) Biotic pump of atmospheric moisture as driver of the hydrological cycle on land. Hydrol Earth Syst Sci 11:1013–1033
Makarieva AM, Gorshkov VG (2010) The biotic pump: condensation, atmospheric dynamics and climate. Int J Water 5:365–385. doi:10.1504/IJW.2010.038729
Makarieva AM, Gorshkov VG, Li B-L (2004a) Ontogenetic growth: models and theory. Ecol Model 176:15–26
Makarieva AM, Gorshkov VG, Li B-L (2004b) Body size, energy consumption and allometric scaling: a new dimension in the diversity-stability debate. Ecol Complex 1:139–175
Makarieva AM, Gorshkov VG, Li B-L, Chown SL, Reich PB, Gavrilov VM (2008) Mean mass-specific metabolic rates are strikingly similar across life's major domains: evidence for life's metabolic optimum. Proc Natl Acad Sci USA 105:16994–16999
Makarieva AM, Gorshkov VG, Sheil D, Nobre AD, Bunyard P, Li B-L (2014) Why does air passage over forest yield more rain? Examining the coupling between rainfall, pressure, and atmospheric moisture content. J Hydrometeorol 15:411–426. doi:10.1175/JHM-D-12-0190.1
Mojzsis SJ, Arrhenius G, McKeegan KD, Harrison TM, Nutman AP, Friend CRL (1996) Evidence of life on Earth before 3,800 million years ago. Nature 384:55–69. doi:10.1038/384055a0
Mokyr J (2013) Is technological progress a thing of the past? VOX (CEPR's policy portal). http://www.voxeu.org/print/9700
Nowak MA, Tarnita CE, Wilson EO (2010) The evolution of eusociality. Nature 466:1057–1062. doi:10.1038/nature09205
Nowak et al (2011) Nature 471:E9–E10. doi:10.1038/nature09836 (Reply)
Peterson C, Park N (2010) What happened to self-actualization? Commentary on Kenrick et al. (2010). Perspect Psychol Sci 5:320–322. doi:10.1177/1745691610369471
Schwartzman DW, Volk T (1989) Biotic enhancement of weathering and habitability of earth. Nature 340:457–460
Sheil D, Murdiyarso D (2009) How forests attract their rain: an examination of a new hypothesis. Bioscience 59:341–347. doi:10.1525/bio.2009.59.4.12
Van der Hejden MG, Horton TR (2009) Socialism in soil? The importance of mycorrhizal fungal networks for facilitation in natural ecosystems. J Ecol 97:1139–1150. doi:10.1111/j.1365-2745.2009.01570.x
Vause WG (1995) The subsidiarity principle in European union law – American federation compared. J Int Law 61–68

Whitman WB, Coleman DC, Wiebe WJ (1998) Prokaryotes: the unseen majority. Proc Natl Acad Sci USA 95:6578–6583

Wilderer PA (2009) Climate change and its consequences for water and wastewater management. Front Environ Sci Eng China 3:2–11. doi:10.1007/s11783-009-0001-z

References of Section 3.4

Aristotle (1998) *The metaphysics* translated with an introduction by H. Lawson-Tancred. Penguin, London

Caney S (2005) Cosmopolitan justice, responsibility, and global climate change. Leiden J Int Law 18:747–775

Rivetti I, Fraschetti S, Lionello P, Zambianchi E, Boero F (2014) Global warming and mass mortalities of benthic invertebrates in the Mediterranean sea. PLoS ONE 9(12), e115655. doi:10.1371/journal.pone.0115655

Scholte J-A (2008) Defining globalisation. World Econ 31:1471–1502

UNHCR (2014) UNHCR global trends 2013. http://www.unhcr.org/5399a14f9.html. Accessed 14 Jan 2015.

References of Section 3.5

Fachinger U (2008) The end of discrimination in social security for the elderly? Z Gerontol Geriatr 41(5):360–373

Gibb H, O'Leary KG (2014) Mercury exposure and health impacts among individuals in the artisanal and small-scale gold mining community: a comprehensive review. Environ Health Perspect 122:667–672

Global Footprint Network (2014). http://www.footprintnetwork.org/de/index.php/GFN/page/world_footprin

Grambow M (2013) Nachhaltige Wasserbewirtschaftung, Konzept und Umsetzung eines vernünftigen Umgangs mit dem Gemeingut Wasser. Springer, Wiesbaden, pp 307–400

Haber W (2010) Die unbequemen Wahrheiten der Ökologie, Eine Nachhaltigkeitsperspektive für das 21 Jahrhundert. Oekom-Verlag, München

Islam N (2013) Towards a sustainable social model: implications for the post-2015 agenda. UN-world economic and social survey (WESS) 2014 background paper

Kilching P (2008) Vorlesung ETH Zürich, FS 2008. http://www.sec.ethz.ch/education/FS2009/Wald_Landschaft/dokumentestud/Subsistenzwirtschaft

Kirmayer LJ (2002) Psychopharmacology in a globalizing world: the use of antidepressants in Japan. Transcult Psychiatry 39:295

Mello AF, Maciel MR, Fossaluza V, de Paula CS, Grassi-Oliveira R, Cavalcante-Nóbrega LP, Cividanes GC, Soussumi Y, Soussumi SP, Perissinotti DNM, Bordin IA, Mello MF, Mari JJ (2014) Exposure to maltreatment and urban violence in children working on the streets in Saõ Paulo, Brazil: factors associated with street work. Rev Bras Psiquiatr 36:191–198

The Global Education Project. http://www.theglobaleducationproject.org/earth/human-conditions.php

Chapter 4
Towards Sustainable Economies

Michael von Hauff, Franz-Theo Gottwald, Katharina Stöckl, Jelena Kurz, and Carolin Böker

4.1 Establishment of a Sustainable Economy Requires a Proper Balance Between Centralized and Dezentralized Structures

Michael von Hauff

4.1.1 Key Messages

In the current sustainability debate with economic justifications, other areas such as sustainable consumption clearly indicate that a decentralized level of execution is required to implement a sustainable development paradigm.

M. von Hauff (✉); Corresponding author for Section 4.1
Technical University of Kaiserslautern, Kaiserslautern, Germany
e-mail: hauff@wiwi.uni-kl.de

F.-T. Gottwald (✉); Corresponding author for Section 4.2
Schweisfurth Foundation, Munich, Germany

Institute for Resource Management, Humboldt University, Berlin, Germany
e-mail: CThomas@schweisfurth-stiftung.de

K. Stöckl (✉); Corresponding author for Section 4.3 • C. Böker (✉); Corresponding author for Section 4.4
Technical University of Munich, Munich, Germany
e-mail: katharina.stoeckl@tum.de; carolin.boeker@tum.de

J. Kurz (✉); Corresponding author for Section 4.3
Bavarian State Ministry of the Environment and Consumer Protection, Munich, Germany
e-mail: kurz.jelena@gmail.com

P.A. Wilderer, M. Grambow (eds.), *Global Stability through Decentralization?*,
Strategies for Sustainability, DOI 10.1007/978-3-319-24358-0_4

4.1.2 Introduction

In 1992, at the United Nations Conference on Environment and Development (UNCED) in Rio de Janeiro, 178 countries made a commitment to the sustainable development paradigm. The global conference resulted in great international popularity for the paradigm of sustainable development and a growing political focus on ensuring its implementation. The guiding action plan for the twenty-first century is called "Agenda 21." At the Rio Conference, a series of further resolutions were prepared such as the Rio Declaration on the Environment and Development (the right to development so as to meet the needs of present and future generations), the Convention on Climate Change (efforts to stabilize greenhouse gas concentrations to prevent interference with climate systems), the Convention on Biological Diversity, and the Statement on Deforestation (management and conservation of the forests according to the principle of sustainability).

Although not legally binding, the adopted documents and conventions have the character of framework conditions for a new international partnership. The Rio Conference was followed by a series of subsequent conferences which further specified the individual topics such as the international Conference on Population and Development in 1994, the World Summit for Social Development in 1995, and the Climate Conference (Kyoto Protocol) in 1997. Again, these resulted only in recommendations and not in any binding resolutions.

In 2002, a follow-up conference was held in Johannesburg. It focused on the implementation plan, which included new targets and programs for environmental protection and poverty reduction. Most recently, the Rio + 20 Conference was once again convened in Rio de Janeiro in 2012. The participants specified and renewed the political will and the movement towards sustainable development. A major topic was the "Green Economy." The many international conferences illustrate the fact that the concept is a global paradigm, and the recommendations and resolutions are acknowledged by all countries.

At the global level, however it was obvious from the start that the agreements were of a highly non-binding nature. Besides the global dimension, it quickly became apparent that a concrete implementation of programs and activities of the sustainability paradigm would be shaped by decentralization. As early as 1997, prior to the Johannesburg Conference, the goal was agreed that all countries should develop national sustainability strategies by the year 2002.

Clearly to an increasing extent, individual states, municipalities (local Agenda 21), corporations (eco-efficiency, corporate social responsibility, etc.) as well as other organizations like churches are prepared to introduce the ideas of sustainable development. The following discussion concentrates on selected examples of decentralization in the area of economics. But first, let's look at the main requirements for sustainable development in terms of the economy.

4.1.3 Requirements for a Sustainable Economy

Sustainable development consists from an economic perspective, in securing the basis of life and the means of production. This definition provides the basis for calling for the global and permanent protection of the environment and, on this basis, to organize and stabilize our economic and social systems. The demands of sustainable development go beyond this and explicitly require intra- and intergenerational equity. Also, in the economic debate there are controversial positions regarding the relationships between the environment and economics as well as social justice. Even today, the relationship among environment and economics is sharply debated by the advocates of neoclassical economics and those who promote ecological economics (Costanza et al. 2001; von Hauff 2014). There are also opposing positions to be heard in the discussions about intra- and intergenerational equity.

A broad international consensus has formed which adheres to the idea today that sustainable development is based on a three dimensional concept. According to this thought the three dimensions, ecological, economic and social, should be of equal importance. It also takes into account the fact that humanity cannot survive without certain conditions of nature or healthy ecologic systems. It follows that an economic or a social system cannot be sustained in isolation. Our long term survival depends on the balanced interaction of economy and society with the ecologic system.

It is already clear from these contextual boundaries of sustainable development that the primacy of the economic dimension must be abandoned. In so doing, the criticism that more and more areas of our life are determined by economic thought and activity can at least be expressed in relative terms. The demand of sustainable development is—as already mentioned—to bring the three dimensions into balance. Of course, this is the ideal condition, which may be pursued but never attained in full measure. We must not only examine each of the three dimensions. It is really about the complementarity of the three dimensions.

Complementarity can be illustrated by an example: Clean air and clean water improve people's health and increase the productivity of human capital. That is to say: Improving the ecological systems (cleaner air and cleaner water) improves the well-being or the quality of life of people (strengthens good health) and increases the efficiency of employees (strengthens productivity of human capital). However, this example of the complementarity of the three dimensions should not obscure the fact that sustainable development is justified differently in the context of the economic debate. This can be most clearly discussed in terms of the substitution rule and the different assessments under the terms of neoclassical economics and environmental economics.

In the context of sustainable development, the issue for proponents of neoclassical economics is that future generations have a supply of capital stock equivalent to that available to generation now living. The aim of this position is to insure that future generations also have the chance to satisfy their needs in the same way as the present generation. The capital stock is cumulative and made up of both natural and real capital.

As a rule, there is an inverse relationship of natural capital to real capital in that the natural capital such as forests, land, as well as rare metals and rare earths and fossil fuels, is increasingly reduced and the real capital is increased. Consequently, consumption of net capital is generally required in order to increase real capital. It is important when substituting real capital for natural capital that the cumulative capital stock remain the same. The reduction in net capital is compensated by the corresponding expansion of real capital. The literature refers to this position as the concept of "weak sustainability."

The substitution rule is fundamentally rejected by the proponents of ecological economics. Their criticism of neoclassic theory essentially involves a restriction on the individual pursuit of profit and utility maximization. In ecological economics, there is an "eco-centric" view, whereby the starting point of all argumentation is the survival and preservation of ecologic systems. In particular, this requires the prevention of irreversibilities in ecosystems, something that neoclassical environmental economics barely mentions.

On the basis of the evolutionary world view of ecological economics characterized by unknowns and uncertainties, price alone is not sufficient as a control function for economic activity. This is why an aggregate capital stock consisting of natural as well as real capital must be preserved (strong sustainability). This is a topic for residence research in that it addresses the limits, but also adaptation to environmental systems (Wilderer and von Hauff 2014). It is becoming more and more evident that the conventional "mainstream economy" has not yet adopted the paradigm of sustainable development to an adequate extent: (Ruth 2006, p. 335ff.)

- Findings of other disciplines are largely negated
- An efficiency paradigm is adhered to (the search for equilibrium, for example, in the form of a market clearance) and
- Spatial and temporal trends as well as cultural and historical context are not adequately taken into account

The controversy between neoclassic environmental economics and ecological economics reveals further polarization when it comes to the issues of environmental sustainability and economic growth. The antipodes are clearly defined: At one end, it is all about realizing that growth and sustainable development go "hand in hand." At the other, growth leads to a heavy burden on the environment and can cause irreversible damage to nature. The first position is characterized by the belief that unlimited growth is possible. The second position acknowledges limits to growth. This is why the classical environmental economists are often referred to as "growth optimists" and the proponents of ecological economics as "growth pessimists."

The concepts of weak and strong sustainability are two standalone paradigms, which are nevertheless sometimes both labeled as inadequate in the literature. Clearly, as the discussion of climate change shows, for example, "*Whether one believes in one paradigm or the other is ultimately just that: A matter of belief*' (Neumayer 1999, p. 41)." The statement refers to the underlying premises of the two positions, where neither weak nor strong sustainability is empirically supported to a sufficient degree. They represent two extreme starting conditions that seem to be based on their own conclusive justifications and the dissociation of their respective positions.

However, there was also a relatively early effort to overcome these opposing views. These efforts resulted in a third paradigm, called "balanced sustainability." The contributors to this concept include Lerch and Nutzinger (1998), Steurer (2001), and Hedinger (2007). Again, these contributions do not provide any uniform argumentation. What they have in common is not an all-inclusive assessment, but rather individual empirical studies of cases, which are then used to argue for or against substitution, as appropriate. A substitution is always possible if the substance of the natural capital is not immediately endangered.

In terms of balanced sustainability, further differentiation has occurred in the recent past. The topic of marine fisheries is often cited as an example. Garmendia and others have reached the following conclusion: "According to the UN Millennium Ecosystem Assessment (2005), the depletion of fish stocks is one of the major examples of potentially irreversible change to an ecosystem that results from present unsustainable practices in marine ecosystems. The World Summit on Sustainable Development also establishes that fish stocks should be recovered to sustainable levels, and sets a deadline of 2015 for reaching the objective of Maximum Sustainable Yield (MSY)." (Garmendia et al. 2010, p. 96) The example clearly shows the necessity of breaking down the sustainable development paradigm to specific areas or to decentralize the issues.

Davis, on the other hand, reaches a broader interpretation in her formulation of the requirements of a balanced sustainability: "The illustrated middle pathway is to protect the environment by commodifying it, or bringing the externalities of the environment and nature into the market. However, there needs to be a movement towards an eco-socio-feminist perspective if we are to gain both social and environmental equality, thereby achieving the ultimate goal of sustainability." (Davis 2013, p. 119) It should be noted that Davis combines the environmental and social dimensions, something that has been largely neglected until now in the discussions of balanced sustainability. One of the initial conclusions about balanced sustainability can be stated as follows: The highest priority in terms of balanced sustainability is a satisfaction of basic needs and improvements in the standard of living for the present and future generations worldwide.

From this, it follows that people, not nature, should take center stage. The advocates of sustainability without explicit limits to growth (for example, a zero growth economy) agree: the critical element of change required to bring harmony to the current conflicting goals, is not the extent, but rather the kind or the direction of economic development. The continuous growth must have an environmentally friendly aspect. In this context, the discussion turns to qualitative or sustainable growth (von Hauff and Jörg 2013). Such growth is achieved by reducing and preventing material and energy input through conservation, reparable goods, recycling, improved efficiency, material substitution, and basic structural change. This harmonization of growth and environmental quality may well lead to a slowdown in growth. This is because an economic and ecologic optimization of the development curve may be different than previously experienced.

However, it must be noted that balanced sustainability is somewhat controversial. Some proponents of ecological economics no longer see strong sustainability and economic growth as being in fundamental opposition, which is the basis for the paradigm of balanced sustainability. In effect, this has already achieved some agreement. The debate is no longer about a fundamental controversy regarding growth targets, but about the following question: Is it possible to have economic growth in a limited world through the use of eco-friendly innovation and technology and the other areas mentioned above like recycling, substitution, and repairable goods? Assuming this is feasible, there are already several measures and concepts that contribute to the harmonization of sustainability and economic development. The following example is intended to illustrate an approach to a sustainable resource strategy and the concept of a sustainable industrial park.

4.1.4 The Concept of a Sustainable Industrial Park

Agenda 21 expressly points out that companies are the major actors in the implementation of sustainable development. In this context, debate over the macroeconomic or social responsibility of the corporation has led to the development of many alternative concepts. One of the concepts that gained attention worldwide is called "Corporate Social Responsibility" (CSR), which was conceived by the European Commission. The CSR concept applies to companies and involves the application-oriented interpretation of the sustainable development paradigm.

The European Commission defines corporate social responsibility as follows: CSR is a concept that serves as a foundation on which corporations can integrate, on a voluntary basis, their activities with social and environmental issues and their inter-relationships with stakeholders. The European Commission views CSR as a business contribution to sustainable development (European Commission 2002). Keep in mind, however, that the majority of companies fall into the small and medium sized company (SME) category. In Germany, for example, 99.5 % of the companies are assigned to the SME category. Although over 70 % of all SMEs recognize the importance of CSR, many of these companies have neither the financial nor the human resources to implement CSR. However, there is the possibility, for example, in the context of an industrial park or industrial zone, for them to initiate joint CSR activities.

Industrial parks have still not been recognized as potential actors in the implementation of sustainable development although such areas could make a large contribution to sustainable development. In fact, they are in position to contribute even more than individual firms. A sustainable industrial park, based on the concept of sustainable development, is a concept that includes more than just the environmental aspects. This approach gives equal consideration to the three dimensions: environmental, economic, and social, and that ideally, means equal consideration of all three dimensions. The concept of a sustainable economic business zone must explicitly address the other two dimensions.

Although the paradigm of sustainable growth has gained much attention at the global, national, regional, and municipal levels and is already being implemented by specific concepts or strategies at the individual enterprise level, the interest in its specific application to entire industrial or commercial areas has just recently begun. The development of the concept is still sailing in uncharted waters. But, as mentioned earlier, many opportunities are only available to small and medium sized companies if performed jointly, not as individual companies. In the next section, many ecologic, economic, and social activities are listed for the purpose of example:

Environmental measures: Development of sustainable water management, to include Ensuring ground water quality, preventing waste water pollution, rain water management in the form of seepage, introduction of joint energy management)

Economic measures: Joint educational and training concepts, joint job bank for internships, joint logistics systems, etc.

Social measures: Community cafeteria, childcare, establishing a fitness center, cultural events, etc.

By differentiating the three kinds of measures it becomes clear that there are overlapping areas. Joint educational and training concepts can make a major contribution to the firm's competitiveness, but this can also convey environmental requirements of the firm and eventually lead to a strengthening of the social connections between the employees working in the industrial park.

4.1.5 Requirements for a Sustainable Resource Strategy

Strengthening resource efficiency is a key area of Germany's national sustainability strategy. There are also many demands for a decentralized implementation when developing a sustainable resource strategy. First, it can be said that the discussion of limited resources is drifting more and more away from fossil fuels to focus on rare earth and metals. One explanation for this is that modern and forward looking production processes and products such as computers, mobile phones, displays, medical technology, renewable energy sources like wind parks and solar cells, energy saving lamps, high performance technologies and even electric mobility depend on rare earth and metals.

These are the technologies of the future. A major area of any sustainable resource strategy is the efficient recycling of such resources. In the previous resource-based economic models, the resources are consumed by manufacturing or direct consumption. Many possibilities already exist today for the recycling of technology metals. Theoretically, it can be assumed that rare metals and rare earth elements can be up to 100 % recycled. This would correspond to the laws of thermodynamics where no metal would be lost.

In reality, however, such complete recycling of metals is not feasible as there is no comprehensive closure of material cycles, partially for technological reasons. The dimensions of this problem can be illustrated by a simple example: In 2008, a total of 1.3 billion mobile telephones were sold. As a result of inadequate collection and recycling processes, 31t gold, 235t silver, 12t palladium, 2.4t indium, and 4.9t cobalt will be lost (Reller and Dießenbacher 2014). A further differentiation of recycling methods is required.

Also, a sustainable resource strategy must take other areas into account, for example, resource extraction and resource utilization, which should be permitted only within the carrying capacity of the ecologic systems. If, for example, during the extraction of resources, it comes to serious and even irreversible damage to the environment, it would be contrary to the requirements of sustainable development. A necessary measure would be to restore the region subsequent to the extraction of the resource. In addition, there should be no negative effects on the health of the local inhabitants of the region or the employees who perform the extraction of the metals.

In the environmental economics debate such considerations are called negative external effects, which are to be eliminated in the framework of the internalization strategies. Such a process is always reactionary and inadequate.

From the discussion it is clear that a sustainable resource strategy must be broken down in many individual areas, that is, decentralization is required. A sustainable resource strategy is also concerned with improving resource efficiency. For example, the resource efficiency of a product can be improved throughout the entire life cycle of a product. This starts with the product design and continues through consumption to its final disposal and recycling.

A review of the existing resource strategies reveals that as a rule they are still characterized by different priorities and special interests. The strategies of the EU member states generally place primary interest in securing competitiveness and innovation capacity of the respective industrial sectors (von Hauff 2014, p. 142). For countries like Germany and Japan, which are dependent on the import of resources, securing supplies is the top priority. On the other hand, countries like China, Russia, and the USA, which do not need to import the resources, can pursue geo-political and security objectives.

A resource strategy must therefore take into account the interactions between actors and the unforeseeable events such as risks and uncertainties. The resource strategy must be constantly updated to account for changing conditions. One of the first steps in the planning and implementation of such a strategy is to formulate and define objectives.

The overall objective is concerned with improving the standard of living in resource importing countries as well as in the exporting countries. This is not only about increasing the income to improve the material quality of life situation. It is also about the intangibles such as the improved standard of health, the observance or attainment of human rights, and the strengthening of equal opportunity. Consequently, a sustainable resource strategy is composed of many combined elements, which explains why implementation requires many responsible actors.

4.1.6 Conclusions

The sustainable development paradigm first went global, that is, it was first decided and introduced by the international community in the twenty-first century. A series of international conferences since 1992 have focused attention and specified individual topics. The program of action known as Agenda 21, in fact, quite specifically shows that decentrally organized concepts and activities are required to implement the paradigm of sustainable development.

From the perspective of sustainable economics, the theoretical foundation of sustainable development includes the opposing positions of weak and strong sustainability. A third rationale is based on balanced sustainability. The fact that there are various approaches in theoretical reasoning means that strategies and measures tend to focus on one approach. It is easy to conclude that no comprehensive resource strategy can be expected. It will require policy legislation.

The explanation for this is that the requirements of sustainable development are not readily prescribed by policy, but must be left to the responsible individual actors (companies, and social organizations like churches, healthcare facilities, as well as schools and universities). This demands a "fundamental change in the way people think," something has not yet occurred in sufficient numbers. The "primacy of economics" still dominates, and such thought still characterizes, to a large extent, the behavior of any people.

4.2 Is Bioeconomy the Road to Decentralization?

Franz-Theo Gottwald

4.2.1 Key Messages

Bioeconomy is the ambitious strategy to implement biotechnologies on a broad political and social basis. Barely noticed by the public, a powerful alliance between industry, investors, politics and science has been set up which aims to turn all life on earth into a commodity which can be freely traded and negotiated. Up to now, no culture of participation is possible when it comes to setting up the political framework or deciding on the future structure of the endeavor. In view of the risks inherent in a higher utilization of plants and reduced use of fossil fuels, a relevant discourse within societies would be paramount.

4.2.2 Bioeconomy: The Economization of Life

The term bioeconomy was introduced as early as 1997 by the geneticists Juan Enriquez-Cabot and Rodrigo Martinez during a meeting of the American Association for the Advancement of Science. In a summary of their contribution about the economic potentials of genomics, Juan Enriquez-Cabot defined bioeconomy as an „economic field which uses novel biological knowledge for commercial and industrial purposes" (Enríquez-Cabot and Martínez 1998, pp. 925 f., translated by the author). This definition reveals that bioeconomy does not constitute the ecological alignment of economics, but the economical alignment of ecology—or in other words—the economization of all living entities (Gottwald and Krätzer 2014, p. 12).

The German Federal Ministry for Education and Research (2010, p. 2) defines bioeconomy as follows: "The concept of bioeconomy covers agricultural economy and all manufacturing sectors and associated service areas that develop, produce, process, handle, or utilize any form of biological resources, such as plants, animals, and microorganisms. This spans numerous sectors, such as agriculture, forestry, horticulture, fisheries and aquaculture, plant and animal breeding, the food and beverage industries, as well as the wood, paper, leather, textile, chemical and pharmaceutical industries, and aspects of the energy sector. Bio-based innovations also provide growth impetus for other traditional sectors, such as in the commodity and food trade, the IT sector, machinery and plant engineering, the automotive industry, environmental technology, construction, and many service industries."

The Federal Government's goal is to use research and innovation to facilitate a structural transition from an oil-based to a bio-based industry, which will also offer much-lauded opportunities for growth and employment. At the same time, research and innovation will be the basis for taking on international responsibility for global nutrition, the supply of commodities and energy from biomass, as well as for climate and environmental protection. This research strategy sets five priorities to continue Germany's path towards a knowledge-based, internationally competitive bioeconomy: global food security, sustainable agricultural production, healthy and safe food, industrial use of renewable resources and biomass-based energy sources (Federal Ministry for Education and Research 2010, p. 2).

On first perusal, the above sounds harmless and environmentally friendly. Only a closer scrutiny reveals the intention of exploiting the biosphere commercially, while employing genetically-modified organisms to implement far-reaching changes in the fields of nutrition, chemistry and pharmaceuticals. This enforced conformity in many areas aims at the total economization of all living organisms: Alterations of the genetic structure of plants and animals as well as genetically engineered medications and therapeutics are means to these ends. Synthetic biology which creates artificial organisms in the lab takes this one step further. This is a very unsettling development, given that the consequences of our present genetic engineering practices are totally unknown and unknowable. In addition there is no way of predicting how artificial life will interact with the environment.

4.2.3 *Bioeconomy and Agricultural Production*

The bioeconomical approach to solving the world hunger problem basically comprises two components: first, a further increase of production via intensifying worldwide agriculture, primarily with biotechnological methods, and, secondly, an increase in efficiency within the scope of waste management and recycling as well as a reduction of food waste (Gottwald and Krätzer 2014, pp. 66ff.). Better waste management and additional processing of waste materials is a long-standing claim by many environmentalists, stakeholders, academics and politicians.

However, declaring a further biotechnological intensification of agriculture as a sustainable solution might be problematic. First, industrial high-tech agriculture is expensive and resource-intensive which from a social and ecological perspective is not sustainable. Over-use of heavy machinery, pesticides and fertilizers has already become a serious problem for biodiversity, residues, soil fertility and water management. Secondly, biotechnological procedures favoured by bioeconomics are not suitable to support poor rural population given that they are capital-intensive, knowledge-based and protected by patent law.

Thirdly, there is sufficient scientific proof, that small-scale, local low-budget agriculture is ideally suited to combat poverty and hunger in rural societies (Gottwald 2015, pp. 261ff.).

Bioeconomy represents a science and capital-intensive, structurally centralized approach. It may be said that at this writing many establishments have been provided with the opportunity for decentralized, biotech-based operations and growth. However, this is done without any regard to the many organic agricultural alternatives which would allow sustainable developments without additional pressure upon soil, water, plants, animals and human beings. Loss of diversity and evolutionary potential will be one price of these limited, bio-industrial solutions. The German Federal Ministry of Education and Research (2014) states that the shift towards a bio-based economy would lead to new forms of centralization as well as to regionalization and decentralization. However, it remains unclear which processes might be decentralized; particularly in view of agricultural production, operating cycles, supply and marketing chains and management systems being oriented towards global structures.

Vertical Farming, for instance, is such a high-tech-approach for the production of „biomass", particularly in urban areas: In practically closed cycling systems, plants and animals are "produced" in gigantic multi-level towers. Plants grow without soil, fed by running water systems enriched with fertilizers and other hydro-cultural systems. Artificial light ensures faster growth—the whole environment is less agriculture than technical production made by engineers. At first glance, the idea sounds conclusive: With its enormous savings regarding water, pesticides and land, while being seasonally independent and safe from weather extremes, droughts and storms, Vertical Farming could be the answer to food shortages in urban centers.

But beside the high energy costs (e.g. Zeidler et al. 2013) due to artificial lighting, many fundamental questions arise from the concept of vertical farming: Do farmers/does the society want to go down this path of food production with all its inherent consequences? What concept of life are those technological approaches based on? And—with reference to the centralization problem—who invests in these concepts and procedures? Who builds vertical farming towers, who controls them, who is responsible for costs and prices, for the distribution of the food? And: If technologies like vertical farming would be implemented as effective measures to combat world hunger—who pays for these cost-intensive technologies? Countries with severe starvation problems lack basic infrastructural facilities. It is doubtful that vertical farming and other comparable technologies can effectively contribute to reducing poverty and hunger in urban and rural areas as well. On the contrary, monopolizing and centralizing tendencies are increasingly shifting the balance of power towards only a few, consistently expanding global companies.

4.2.4 Aggressive Lobbying: Disproportional Funding

The promotion of the bio-economic strategy by industries, science and the political administration is ultimately a re-interpretation of the concept of sustainability for the benefit of technological solutions to major corporations. Bioeconomy is described as sustainable, progressive, innovative and indispensable for the good of all humanity. Critics have been silenced. This happens—increasingly in the media—by defaming sceptics as reactionary, ideologically blinded or overprotective.

Civic engagement is annoying for the whole biotech industry. The former CEO of BASF, Jürgen Hambrecht, expresses his views bluntly: "Europe must not miss the opportunities of green biotechnology. We must not be guided by irrational fears. There is not a single scientific proof that plant biotechnology is harmful to the environment or human beings. On the contrary, we will open up many doors to better health and quality of life! (...) Europe must remain a driver of innovation! (...) But above all, I wish that politics explicitly acknowledges new technologies. A mere moderation of public opinion is not enough"(Hambrecht 2010, p. 27). However, skepticism is indicated when considering the hard facts: Risk research? Technology assessment? None in practice. It should be legitimate for a democratic society to demand accountability, the opportunity to recall new technologies and forestall any of their consequences. Finally, it is society that ultimately has to bear the costs.

Policy fosters bioeconomic strategies and grants billions of tax revenues to industrial stakeholders. The German Federal Government is very generous when it comes to the implementation of bioeconomy. The High-Tech Strategy for Germany was supported with 27 billion euros between 2010 and 2013. In addition, 2.4 billion euros were invested by the BMBF for the concept of Bioeconomy 2030. Problematic with the governmental funding practice is the close connection of state institutions and industry and the missing democratic legitimacy of these payments is problematic with this kind of governmental funding practice. For instance, there is a public-private partnership between the EU and an industry group, provided with nearly four billion euros from 2014 to 2020, including one billion from the EU (Bio-based Industries 2014). The Federal Government supports companies in the implementation of a high-tech strategy, while other alternative forms, whose sustainability and benefits to the public have been proven, are simply disregarded. The Federal Organic Farming Scheme (BÖLN), for instance, which is part of the sustainability strategy of the Federal Government and the Federal Ministry of Food and Agriculture (BMEL) is chronically underfinanced. The BÖLN was set up to conduct science in organic agriculture and promote sustainable forms of agricultural land use. Target groups of the various projects, trainings and information measures of BÖLN are not only producers, processors and trade, but also consumers, teachers, the media and other multipliers. This is precisely why this program is also socially and ecologically relevant. However in recent years, funding for the BÖLN has been further cut: while in 2003 35 million euros a year flowed into the program, the annual budget today amounts to only 16 million euros (Gottwald and Krätzer 2014, pp. 145f.).

Due to this preference for Bioeconomy there is a dearth of funds for research into organic forms of agriculture as well as alternative forms of decentralized economic and manufacturing activities. This is particularly deplorable as there are a number of model approaches: Blue Economy for instance has developed a fish farming and greenhouse system which utilizes fish feces as fertilizer for the plants. These plants in turn filter the water, thus keeping the fish healthy. These alternative and sustainable ideas may sound idealistic, but they are less risky and more responsible, and also protect nature and the environment much more so than Bioeconomy. There are myriad alternatives; in addition to the above mentioned example there are solutions of flexible regional networks operating within closed loop material cycles, or locally adjusted systems of self-sufficiency with regard to regenerative energy supplies, food production and a food and agricultural economy organized along self-sufficiency lines. This also includes local, national or global networking in the fields of knowledge, technologies and finances—which represents real decentralization!

A policy—in contrast to Bioeconomy policies—in line with the guiding principles of increasing sustainability through efficiency, consistency and sufficiency, which furthers research and economic activities in keeping with the principles of precaution, responsibility, intergenerational fairness as well as biodiversity, is possible. Divers political steering mechanisms are in place and range from regulatory laws, to tax laws all the way to planning laws. Failure to implement alternatives to Bioeconomy would constitute dire political negligence.

4.3 Sustainable Economy in a Globalized World: Models and Solutions to Induce Sustainable Development

Katharina Stöckl and Jelena Kurz

4.3.1 Key Messages

As a result of the continuous expansion of the international trade, an unbroken trend towards globalization and a concentration of economic power on global players is obvious. This tendency implicates serious risks for the long-term orientation of global business and for the establishment of a sustainable economy within the global ecological capacity. Therefore, substantial changes in current economic structures are necessary to induce a transformation towards a more sustainable economic development. Decentralized structures as well as central governance systems—which need to be promoted by different actors and legal frameworks—will help to establish a more resilient economic system, which focusses on the benefit of the global society as a whole under a social and ecological perspective.

4.3.2 Introduction

For a long time, enhancement of economic growth and profit-driven productivity was seen as the main priority in economy (Douthwaite 1993). As a result of alarming environmental events as well as the increasing public awareness on the depletion on natural resources and on environmental destruction (e.g. Brown 1990; Hempel 1996) the need for rethinking development and economics was recognized on a global scale. This finally resulted in several international conferences and claims for action—the first of their kind were the pathbreaking Brundtland-Report "Our Common Future" from the World Commission on Environment and

Development in 1987 and the United Nations Conference on Environment and Development in Rio de Janeiro 1992—that shifted the public debate towards sustainable development. According to the Brundtland Report (WCED 1987), the term sustainable development is defined as "...development that meets the needs of the present without compromising the ability of future generations to meet their own needs. It contains within it two key concepts: the concept of needs, in particular the essential needs of the world's poor, to which overriding priority should be given; and the idea of limitations imposed by the state of technology and social organization on the environment's ability to meet present and future needs." Although there are various interpretations and definitions of this term (Haque 1999), a broad international agreement exists on the three dimensional character of sustainable development, including environmental, economic as well as social sustainability.

In terms of sustainable economy, the ongoing trend towards globalization and a shift to centralized structures is often viewed highly critical (e.g. Hirst et al. 2009). While the worldwide financial market or market liberalization has brought many benefits, critics warn of new risks arising by this development.

In the scope of the focus group session "Sustainable Economy", members of the International Experts Group on Earth System Preservation (IESP) discussed the potential threats of the observed tendency towards an even greater globalized market and proposed some ideas and models that may contribute to a more stable and resilient economy. In particular, the importance of decentralized vs. centralized structures for a sustainable development was critically discussed. The results are summarized below, if not otherwise identified.

4.3.3 Current Issues

Regardless of having two categories of companies such as small and medium sized enterprises (SMEs) and multinational enterprises (MNEs), there is an unbroken trend towards globalization. A fundamental part of globalization is the growth of international trade and production. According to Shangquan (2000), "[e]conomic globalization refers to the increasing interdependence of world economies as a result of the growing scale of cross-border trade of commodities and services, flow of international capital and wide and rapid spread of technologies." Results are a continuous expansion of international business arrangements within a highly competitive market characterized by high volatility and rising economies of scale. Moreover, a concentration of economic power on global players under a short term profit pressure can be observed today. Even though international trade is much more facilitated nowadays, SMEs often suffer from increased competition which threatens their long-term viability. This all causes a compulsion of growth and commonly outsourcing of labor, goods and materials etc., which leads to further consolidation and centralization of business structures on a global scale. It is also argued that MNEs influence local and national governments e.g. by threatening them with market withdrawal (Boundless 2014). In the end this development implicates serious risks for the current and future sustainability of economy. Substantial changes in economic development need to be initiated and directed towards sufficiency. Although influencing the current economic trend is a highly complex and challenging task, the focus group discussed the ideas and models following below, which offer solutions to induce a transformation to a more sustainable economic system. Undoubtedly, sustainable development cannot be achieved without the support of different actors including policies and media.

4.3.4 Models and Solutions for a Sustainable Economic Development

4.3.4.1 Establishment/Support of New or Novel Economic Company Structures

A series of company structures that are based on a decentralized organization have already proven to be successful in terms of stability and thus serve us as positive examples. For instance, companies organized as cooperatives are considered to have a higher level of entrepreneurial sustainability and higher resilience (Sanchez-Bajo and Roelants 2011). This became particularly obvious during the financial crisis in 2008, when most conventional banks experienced huge losses, while co-operative banks were largely unaffected by the precarious situation on the financial markets.

The multi-stakeholder governance model as a distinct form of cooperative has become increasingly popular in Europe (Kerlin 2006). The principle of this company structure is that two or more classes of members, e.g. producers, workers and consumers have joined together to implement solutions for common goals. The stakeholders cooperate and participate in the decision-making process, and thus are involved in governance processes. Although several researchers predicted that organizations based on this governance structure will fail, empirical evidence suggests that multi-stakeholder cooperatives are able to successfully govern their companies (reviewed in Leviten-Reid and Fairbairn 2011). "Solidarity (multi-stakeholder) cooperatives represent a rearticulation of the linkages between economic and social spheres in an environment where the global economy and new technologies call for a potentially unlimited mobility of capital, labour and knowledge. The local roots of solidarity cooperatives, which are owned and operated by local actors for the benefit of their members, represent an obstacle to this de-localization and maintain the balance between local socio-economic needs and the challenges and opportunities presented by the local economic system." (Jean-Pierre Girard, Canadian expert on multi-stakeholder cooperatives).

4.3.4.2 Change of Corporate Governance: Reformation of Corporation Laws

A crucial piece of the transformation process towards a sustainable economic development is also seen in a reform of the legal frameworks for corporate governance. Conventional corporate laws are focused on maximizing shareholder profits and short-term gains. This shareholder primacy drive often hinders companies to act in an environmentally and socially responsible way. There has to be a shift away from short-term thinking towards a long-term sustainable focus while management and liability notably have to remain in one hand.

4.3.4.3 Establishment of a Functional Emissions Trading System

The reduction of greenhouse gases to mitigate climate-change induced threats and impacts remains one of the biggest challenges for humankind. As part of national strategies to combat climate change, several countries introduced Emission Trading Systems, such as the European Union, the Northeast and Mid-Atlantic States of the U.S. (RGGI) and New Zealand in order to reduce the emission of greenhouse gases cost-effectively. The European Union Emission Trading System (EU-ETS) involves 31 countries with different abatement costs and has served as a role model for other systems around the world (BMU 2013).

The principle of this market-based approach is that companies have to buy emission permits from an authority—usually from the government—allowing them to discharge a certain amount of a specified pollutant. The permits can be traded on a secondary market. The focus group agreed that the existing Emission Trading Systems in their current forms suffer from major weaknesses and need to be revised in order to successfully contribute to a reduction of emissions. In particular, an oversupply of emission permits hampers the efforts to cut greenhouse gases. Thus, it is strongly recommended to reduce emission permits gradually.

4.3.4.4 Stepwise Reduction of Privileges of Energy Intensive Companies

Several EU policy initiatives have been put into practice to promote energy efficiency and the use of renewable energy sources and technologies. For example, several European member states introduced taxes and market-orientated instruments to induce sustainable energy consumption and production. In order to guarantee competitiveness on the global market, energy intensive industries are often exempt from payments. The current models have disadvantages for small-and medium sized companies or other non-privileged companies and thus lead to an imbalance in competition. Furthermore, it allows large companies to even increase their energy consumption at the expense of SMEs and customers. Yet, there should be a revision of the current exemption regulations and a stepwise reduction of privileges for energy intensive companies.

4.3.4.5 Protection of Common Goods by Decentral and Central Solutions: Limits of Privatization (e.g. Health Sector, Railway Infrastructure)

In recent years, the on-going liberalization of the global trade and investment has led to an increasing growth of the private sector. Private sector participation in the supply of common goods such as water, transportation, energy as well as health, education and media etc. is discussed highly controversial and opponents of privatization argue that it leads to inequality, democratic accountability as well as to environmental depletion. Among the focus group members, a broad consensus existed on the protection of common goods as one of the major challenges in the next years. A combination of both decentral and central strategies may be applied to guarantee the provision of the common goods to the civil society: privatization has to be limited and provision of common goods has to be treated as a state duty. Strong regulations may be established both on a regional as well as on a national scale.

4.3.4.6 Support of Sustainable Consumer Behavior and Consciousness

Another prerequisite for sustainable economy is a change in the current consumer behaviour and consciousness in consumption habits. In societies of most countries, life-style and values are based on the "…spirit of consumerism, implying a direct correlation between maximum consumption and maximum happiness" (Itty 1984; in: Haque 1999). As discussed in the focus group, a combination of different instruments including information, education as well as labelling and new standards for producers is necessary to induce a change in the way people think and to support sustainable consumer behaviour.

4.3.4.7 Establishment of Higher Transparency and Bundling in the Field of Consumption Labels

To date, a broad range of different consumption labels and certificates are on the market for e.g. nutrition or goods covering diverse sustainability issues such as seasonal, fair trade or low energy consumption. The intention of most (state) labels is to encourage consumers to change their consumption habits and to decide for "green" products or services in order to influence the production side up to more sustainability. Since consumers are confronted with multiple different labeling systems with each having their own standards, labels rather lead to confusion instead of informing the consumer sufficiently. Furthermore, little information on labeling system standards suggests that labels are not independently verified, which, in turn, weakens public confidence in labeling. It is thus suggested to establish a more transparent framework that bundles consumption labels, finally resulting in a reduction of label systems. Both global and local labels should be combined in a complementary way.

4.4 Thoughts from a Non-economical Point of View

Carolin Böker

Facing the Earth's crises including global warming, climate change as well as societal and economic instability, humankind must solve very complex and intertwined problems. Planet Earth has always been resilient to destructive events, such as natural catastrophes. In our times the frequency and intensity of these events seem to increase. Hence, we have to adapt even more quickly and find solutions to keep the balance between using and sharing our planet's resources sustainably, socially equitable and keeping its adaptive potential (ergo ecosystem functioning) at the same time. We need central institutions to regulate and supervise the extent of the threats to our planet caused by humankind itself (pollution of soil, air and water; destructive exploitation of forests, land and aquatic systems). Furthermore, these institutions have to manage general governmental and economic matters. In contrast to that, social problems (e. g. living and working conditions, supply of food, drinking water and energy, plus the access to health care and education), which often require quick decisions and actions, have to be dealt by local institutions with low bureaucratic barriers. Nevertheless from case to case it is important that a more centralized organization oversees these local institutions. Sustainable infrastructure, mobility and logistic have to be managed both on a global and decentral level at the same time, as we have to share and allocate resources, goods and services both: worldwide and regionally. In summary we need global and regional organs cooperating together to deal with the earth crisis and to pass on a planet suitable to live on for future generations. I wonder if institutions function is an adequate model for solving problems of mankind. Functioning ecosystems do not have and need a central regulation—human systems urgently do so in a central in decentral way at the same time.

The discussions during the Wildbad Kreuth Workshop on global stability through decentralization confirmed my first thoughts about the topic. In the focus group working on humane living and working conditions, one issue was the need for balance: the balance between human needs for resources and ecosystem functionality as well as the balance between social justice and individual freedom. This balancing might be achieved by a *good* governance, what in turn raises the question: what is *good* governance? Respecting basic human values and their diversity between different cultures and regions should be a central aim of *good* governance on the one hand. On the other hand, we also have to respect the needs of the ecosystems we are living in. Mankind strongly depends on ecosystem functioning, beginning with drinking water treatment (from surface- to groundwater) by functionally different soil layers inhabited by variable microbial communities, and not ending up with comfortable climate zones in certain parts of the world due to ocean streams such as the Gulf Stream. The mentioned balance has to be negotiated over and over again—the great disadvantage for global policies is that a “try and error” tactic is no more possible.

Talking about ethics means to almost all people talking about human ethics, but what is urgently needed are ethics towards nature. In my opinion this is the only way to preserve ecosystem functioning and consequently ecosystem services. The latter should not be mixed up with human use, it means much more providing services to adjacent ecosystems not services to mankind. If mankind does not learn to integrate in nature, like many ancient people did, and accept a limit of population growth, we probably have to face a big bang... sooner or later. This big bang does not include wars for fossil fuels but for food, drinking water and a place to live in. In that case, technical solutions will not help us anymore and no one wants to imagine what happens after that *big bang*. Nevertheless, at the present time there is a way how to attenuate the earth crisis—if people open their mind for non-economical values.

References

References of Section 4.1

Costanza R, Cumberland J, Daly H, Goodland R, Norgaard R (2001) An introduction to ecological economics. Stuttgart, Lucius & Lucius

Davis GR (2013) Appraising weak and strong sustainability: searching for a middle ground. J Sustain Dev 10(1):111–124

European Commission (2002) Mitteilung der Kommission betreffend die soziale Verantwortung der Unternehmen: Ein Unternehmensbeitrag zur nachhaltigen Entwicklung. KOM (2002) 347 endgültig, Brüssel 02 July 2002

Garmendia E, Prellezo R, Murillas A, Escapa M, Gallastegui M (2010) Weak and strong sustainability assessment in fisheries. Ecol Econ 96

Hedinger W (2007) The conceptual strength of weak sustainability. Paper
Lerch A, Nutzinger HG (1998) Nachhaltigkeit. Methodische Probleme der Wirtschaftsethik, Zeitschrift für Evangelische Ethik 42
Neumayer E (1999) Global warming – discounting is not the issue but substitutability is. Energy Policy 27(1):33–43
Reller A, Dießenbacher J (2014) Reichen die Ressourcen für unseren Lebensstil? Wie Ressourcenstrategie vom Stoffverbrauch zum Stoffgebrauch führt. In: von Hauff M (Hrsg) Nachhaltige Entwicklung – Aus der Perspektive verschiedener Disziplinen, Baden-Baden, pp 89–114
Ruth M (2006) A quest for the economics of sustainability and the sustainability of Economics. Ecological Economics 56(3):332–342
Steurer R (2001) Paradigmen der Nachhaltigkeit. Zeitschrift für Umweltpolitik (ZfU) 24:537–566, Frankfurt
UN Millennium Ecosystem Assessment (2005) UN Millennium Ecosystem Assessment. Island Press, Washington, DC
von Hauff M (2014) Nachhaltige Entwicklung. Nomos Publ., Baden-Baden, Germany
von Hauff M, Jörg A (2013) Nachhaltiges Wachstum. Oldenburg Publ. Munich, Germany
Wilderer P, von Hauff M (2014) Nachhaltige Entwicklung durch Resilienzsteigerung. In: von Hauff M (Hrsg) Nachhaltige Entwicklung – Aus der Perspektive verschiedener Disziplinen. Baden-Baden, pp 17–40

References of Section 4.2

Bio-based Industries (2014) A major public and private effort. http://www.bbi-europe.eu/about/about-bbi. Accessed 16 Dec 2014
Enríquez-Cabot J, Martínez R (1998) Genomics and the world's economy. Sci Mag 281:925–926
Federal Ministry of Education and Research (2010) National research strategy bioeconomy 2030. Our Route towards a biobased economy. http://www.bmbf.de/pub/Natinal_Research_Strategy_BioEconomy_2030.pdf. Accessed 20 Nov 2014
Federal Ministry of Education and Research (2014) Bioökonomie als gesellschaftlicher Wandel. Konzept zur Förderung sozial- und wirtschaftswissenschaftlicher Forschung für die Bioökonomie. Referat Bioökonomie, Berlin. http://www.bmbf.de/pub/Biooekonomie-Gesellschaft.pdf. Accessed 21 Nov 2014
Gottwald F-T (2015) Irrweg Bioökonomie. Über die zunehmende Kommerzialisierung des Lebens. In: Agrarbündnis e.V (ed) Der Kritische Agrarbericht. AbL Bauernblatt Verlag, Hamm, pp 259–264
Gottwald F-T, Krätzer A (2014) Irrweg Bioökonomie. Kritik an einem totalitären Ansatz. Edition Unseld, Suhrkamp, Berlin
Hambrecht J (2010) Ohne Risiko kein Fortschritt. Interview mit Frank Stäudner. Wirtschaft & Wissenschaft 1/2010:27
Zeidler C, Schubert D, Vrakking V (2013) CE-study: vertical farm Eden. http://www.researchgate.net/publication/259899768_Feasibility_Study_Vertical_Farm_EDEN. Accessed 20 Nov 2014

References of Section 4.3

BMU (2013) Towards a global carbon market – prospects for emission trading. 11 & 12 Apr 2013 – Documentation and summary, pp 5–6. Federal Ministry for the Environment, Nature

Conservation and Nuclear Safety (BMU). Retrieved 29 Jan 2015 from http://www.bmub.bund.de/fileadmin/Daten_BMU/Pools/Broschueren/ets_conference_brochure_bf.pdf
Boundless (2014) Multinational corporations. Boundless Sociology, Boundless, 14 Nov 2014. Retrieved 28 Jan 2015 from https://www.boundless.com/sociology/textbooks/boundless-sociology-textbook/global-stratification-and-inequality-8/stratification-in-the-world-system-69/multinational-corporations-407-3428
Brown LR (1990) The illusion of progress. In: State of the world. Worldwatch Institute. W.W. Norton & Company, New York
Douthwaite R (1993) The growth illusion: how economic growth has enriched the few, impoverished the many, and endangered the planet. Council Oak Books, Tulsa
Haque MS (1999) The fate of sustainable development under the neoliberal regimes in developing countries. Int Polit Sci Rev 20:199–222
Hempel LC (1996) Environmental governance: the global challenge. Island Press, Washington, DC
Hirst PQ, Thompson G, Bromley S (2009) Globalization in question. Polity, Cambridge, pp 68–69
Itty CI (1984) Development goals and values. In: Itty CI (ed) Searching for Asian paradigms. Asian Cultural Forum on Development, Bangkok
Kerlin JA (2006) Social enterprise in the United States and Europe: understanding and learning from the differences. Voluntas 17(3):246–262
Leviten-Reid C, Fairbairn B (2011) Multistakeholder governance in co-operative organizations: towards a new framework for research. Can J Non-Profit Soc Econ Res 2(2):25–36
Sanchez-Bajo C, Roelants B (2011) Capital and the Debt Trap. Learning from cooperatives in the global crisis. Palgrave MacMillan, New York
Shangquan G (2000) Economic globalization: trends, risks and risk prevention. Economic & social affairs, CDP background paper (1)
World Commission on Environment and Development (WCED) (1987) Our common future. Oxford University Press, Oxford, p 43

Chapter 5
Sanitary Engineering: Central or Decentral Solutions?

Peter A. Wilderer, Martin Grambow, Asher Brenner, and Werner P. Bauer

5.1 Urban Water Management Requires Both Central and Decentral Solutions

Peter A. Wilderer and Martin Grambow

5.1.1 Key Messages

To minimize volume and rate of abstraction from aquifers and man-made reservoirs the century old concept of water supply and sanitation needs to be complemented by innovation concerning use, transportation and pricing of water and wastewater. Particularly in water scarce areas maintenance of assets, cascading use of water and water reuse deserves specific attendance. Wastewater has to be considered a source of water useable as substitute of fresh water. When choosing the water reuse option it might be useful to place the facility where wastewater is converted into useable

The original version of this chapter was revised. The erratum to this chapter is available at DOI 10.1007/978-3-319-24358-0_8.

P.A. Wilderer (✉); Corresponding author for Section 5.1
Institute for Earth System Preservation (IESP), Technical University of Munich, Garching, Germany
e-mail: peter.wilderer@mytum.de

M. Grambow (✉); Corresponding author for Section 5.1
Bavarian State Ministry of the Environment and Consumer Protection, Munich, Germany
e-mail: martin.grambow@stmuv.bayern.de

A. Brenner (✉); Corresponding author for Section 5.2
Ben-Gurion University of the Negev, Be'er-Sheva, Israel
e-mail: brenner@bgu.ac.il

W.P. Bauer (✉); Corresponding author for Section 5.3
ia-GmbH, Munich, Germany
e-mail: bauer@ia-gmbh.de

P.A. Wilderer, M. Grambow (eds.), *Global Stability through Decentralization?*,
Strategies for Sustainability, DOI 10.1007/978-3-319-24358-0_5

water close to the user of the purified water (decentral solution) instead of pumping the water over long distances from central wastewater treatment plants.

5.1.2 Introduction

Availability of water in sufficient quantity and quality is one of the most important preconditions of economic prosperity, public health and societal wellbeing. It is the backbone of sustainable urban water management. However, water supply is just the one side of the coin of urban sustainability. The other but equally important side of this coin represents purification, direct or indirect reuse and safe discharge of the used water. Advanced wastewater treatment is crucial for maintaining the basic function of aquatic ecosystems receiving wastewater discharges. Healthy ecosystems are known to secure life enabling conditions on earth (see Sect. 3.2, Makarieva et al.).

The problems resulting from over-abstraction of groundwater and from the discharge in insufficiently purified wastewater are very well documented in literature. Contemporary scientific knowledge has led to a well established concept of sound management of local and transboundary water systems (Grambow 2008, 2013) and to the development of the technical means required to protect the ecological function surface water ecosystems. Nevertheless, in many parts of the world the water systems are in a catastrophic shape. Presumably, more fresh water is "consumed" by pollution than by using it for drinking, irrigation and industrial production.

As stated in §3 of the Zugspitze Declaration (Anonymous 2008) deeds must follow our words to overcome water related problems. Sustainable water management is a most promising measure to satisfy the food and water demand in the world, thus solving global and local water and food deficiency crises (§7 of the Zugspitze Declaration).

Sustainability is an expression which is not unknown in the water sector. For instance, the concept of Integrated Water Resources Management—often referred to as IWRM—(Anonymous 2000a; Grambow 2013) is an application of guidelines which were set by the United Nations at the conference in Rio de Janeiro in 1993 (Agenda 21, 1993). In parallel, the Water Framework Directive was issued by the Commission of the European Union (Anonymous 2000b). It became the guideline of water management activities in Europe and beyond. It is advisable to design and implement comparable management methods to combat the four world crises highlighted in Fig. 5.1, namely shortage of water, food, energy and jobs in urban and even more in rural areas of our planet (1), Pollution of the atmposphere, soil and terrestrial and marine waters (2), Loss of ecosystem function and of ecosystems as a whole (3) and destabilization of societies and economies (4).

The increase of the population density and an insatiable desire of people for comfort and high-end lifestyle has already caused in many parts of the world water shortage situations. Long lasting drought conditions amplify the deficit in supplying water to people, industry and agriculture. Water supply deficits almost inevitably is followed by deficits in food supply—particularly in low income parts of the

Fig. 5.1 The 4 major threats aggravating the current Earth crisis

world—and subsequently leads to tremendous economic problems and painful social inequality crises. The interrelationship between all those crises is commonly understood but the contemporary political debate is almost exceptionally focused on the single aspect of global warming and the resulting climate change. It appears that CO_2 emission caused by excessive combustion of fossil fuels is considered as the key problem. Solving the CO_2 emission problem might eventually stop global warming but the billions of people on Earth depending on proper supply of water, food and commodities would remain in unchanged miserable conditions. What we need, in addition to climate regulating measures, is an innovative approach to water management and water technology. Considering the complex nature of the water sector, development of innovative tools to facilitate decision making processes in a world of complexity and heterogeneity is an urgent task.

5.1.3 Water Re-Use, an Option to the Overcoming of Water Shortages in Urban Areas

The future development of urban water technology will be influenced by the tremendous water shortcomings within the fast growing cities of the world. Simultaneously with the growth of urban areas the demand for water and food increases—often beyond the capacity of the locally available water resources (aquifers, rivers,

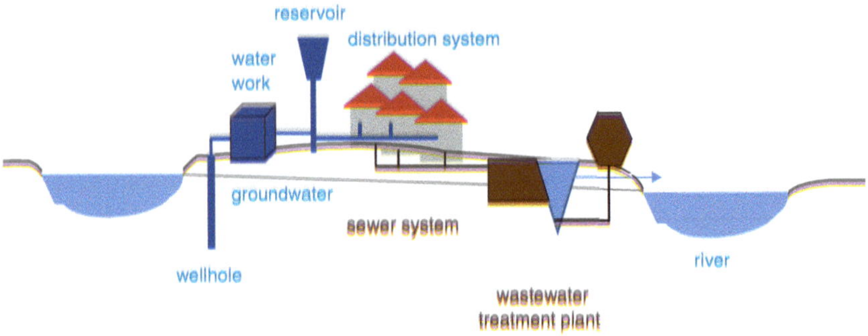

Fig. 5.2 Traditional, single-pass urban water concept

lakes, reservoirs). These resources are rather shrinking in size and quality because of over-abstraction of groundwater (Rodell et al. 2009; Taylor 2014), pollution of rivers and lakes and in coastal zones by intrusion of seawater. Industries, power plants and agriculture compete for the available water which aggravates water supply in urban areas.

Decision makers should understand that copying the technology which has developed over hundreds of years in the water-rich areas of the world makes only limited sense. Instead, tailored solutions are to be implemented, solutions which are adjusted to the local conditions in transition (Wilderer 2007, 2009).

To be able to supply people, industry and farmers with safe water, and to be able to maintain a reasonable public health status with economic prosperity and social stability, we have to consider the use of unconventional sources of water, including wastewater. The cascading use and reuse of water in cities must be seriously taken into account. Decision makers of political institutions as well as private users, enterprises and representatives of industry are to be prepared for a shift from the traditional concept of single-pass water usage (Fig. 5.2) towards the concepts of recovery and re-use. All of us must understand that wastewater is not a nuisance but a valuable resource.

Over-abstraction of groundwater is unknown in ecosystems. To satisfy the water demand of the wide variety of individuals (plants, animals, bacteria) making up a forest ecosystem the concept of cascading use and temporary storage of water is applied by nature. Water is stored in plant and animal tissues, extracellular polymers of bacteria and fungi, and in soil. The stored water is reused in the process of material cycle (see Makarieva, Sect. 3.3). This is in sharp contrast to common practices in cities, industry and agriculture where water is mostly used just once, and discharged thereafter

In order to approach sustainable water managements it is imperative that the rate of water abstraction from natural freshwater resources gets reduced. This is necessary not only to permit long-lasting, satisfactory water supply of people, industry and agriculture, but to maintain the life supporting function of terrestrial and aquatic eco-systems as well. To minimize volume and rate of abstraction some elementary improvements are recommendable concerning water usage, water transportation and pricing. Efficiency is the aim, if this is not enough, sufficiency is the answer. As described by Wilderer et al. (2013) the following aspects might be helpful in the process of identifying solutions:

Agriculture: In many parts of the world only a fraction of the crops grown on agricultural fields makes it to the consumers (Asit Biswas, personal communication). A significant fraction is rotting on its way from the field to the consumers because of the lack of proper transportation and cooling facilities. Any degree of improvement on this side would help overcoming the food supply shortages, and decrease the water demand of agriculture.

Secondly, transformation to water saving agricultural practices would reduce the water consumption of agriculture even further. Development and deployment of innovative concepts such as roof-top farming (growth of vegetable on top of high-rise buildings) is to be encouraged (Kauffman and Bailkey 2000; Wilson 2005).

Municipalities: Because poor maintenance of water distribution systems the loss of water caused by leakages is high, sometimes extremely high. Fixing leaks would diminish local water scarcity problems significantly. Metering water consumption and pricing based on the local income situation would further help decreasing excess use of water.

To further overcome water shortages in cities it is recommendable to consider cascading use of water and re-use practices—at least in new urban development areas. Figure 5.3 provides a graphical representation of the three, currently widely discussed and partially already applied approaches towards water reuse in urban areas:

1. Return of water after advanced purification to the reservoir or aquifer feeding the central water works (Tortajada 2006)

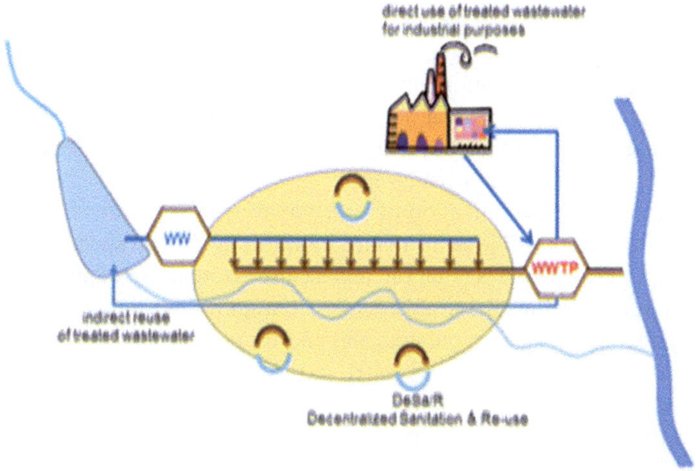

Fig. 5.3 A closer look to options for cascading use of water, and for direct and indirect water re-use in urban areas

2. Direct reuse of urban wastewater in industry purified according the required standards of the receiving industrial plant
3. Direct reuse of water, nutrients and bio-solids in decentralized units after source separation of urine, feces and grey water (Larsen and Gujer 2000; Wilderer 2004; Cornel 2007)

To be able to draw decisions which meet the needs of nature, agriculture, industry and urban dwellers alike, it is necessary to enter into a trans-disciplinary dialogue at the earliest stages of regional, urban and architectural planning (Wilderer 2005, 2014). In addition, transfer of well documented and interpreted information to the end-users is indispensable, but also support of law makers, investors and the press media with solid information. As mentioned above, city planners, architects, investors and representatives of water authorities should understand that copying the technology which has developed over hundreds of years in areas different in climate, economic and educational capacity are not universally applicable. Methods tailored to the specific situation at the site in need for solutions are definitely superior over methods describen in textbooks.

First of all, however, water professionals and water authorities should abandon the sectoral approach which has prevailed throughout the world for centuries. Water supply and sanitation (including urban drainage) are not separate issues but inextricably linked. As mentioned above they resemble the "two sides of the same coin".

Closing the water cycle in an urban setting reduces the stress imposed to the regional hydrology. However, closing the loop bears serious risks with respect to public health. It is unavoidable that chemical substances accumulate in water kept in circulation. Components formerly rated as trace substances are already present in enhanced concentrations, specifically in areas downstream major rivers where water is unintentionally re-used. Of particular concern are nowadays pharmaceuticals, hormones, pesticides, heavy metals and nano-particles. As in earlier times when modern water supply and sanitation system were invented it is of primary importance that pathogenic organisms and viruses are removed from water supplied for potable use. The rapid development of membrane technology makes it possible to control, better than ever before, the spread of water-borne pathogenic organisms and viruses. Further technological developments are necessary to solve drawbacks still associated with membrane systems such as fouling, scaling, leaking, energy consumption and last but not least costs.

With respect to decentralization of water treatment and water reuse, development, testing and responsible performance monitoring of the respective technology is required. Decentralized sanitation and re-use systems (DeSa/R) have to be particularly robust, reliable and affordable. Continuous education and training of design engineers, operators, regulators and end-users is a must.

To make sure that decentralized water and wastewater treatment systems work reliably and cost efficiently, it is necessary to develop robust sensors to provide electronic information about the performance of the dispersed treatment plants (Fig. 5.4). The signals might be electronically or via net of telecommunication transferred to a central monitoring station where advanced signal processing and

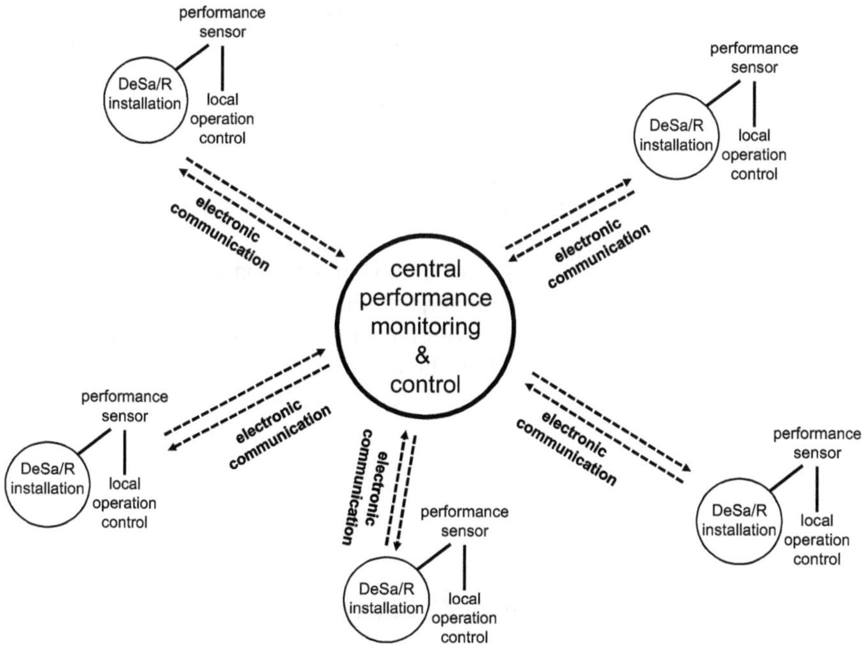

Fig. 5.4 Centralized monitoring and control of decentralized Sanitation and Reuse plants (DeSa/R) on the basis of an innovative sensor and control concept using electronic transmission of performance data

innovative control engineering is performed. The results are to be displayed in a visual form readily understandable by even low-profile service staff.

In conclusion, the need for research and development in the field of water technology is enormous but often under-estimated by the respective funding institutions. Research and technology efforts are to be encouraged. Successful mitigation of water stress in urban and rural areas, in industry and agriculture depends on advances of science, technology, management practices, legal frameworks and public perception.

5.1.4 Livable Cities Need a Livable Hinterland

Even so urban areas pose a big challenge—not only with respect to water management—the rural area surrounding cities, the "hinterland", requires special attention and sustainable development as well. According to the law of symbiosis a city cannot survive without a healthy environment beyond its outskirts. In this context, the term "healthy environment" stands for a balanced interrelationship between human activities and factors enabling nature to perform its generic function. Even at relatively small scale it is important to give room to well-functioning ecosystems, to forests and wetlands in particular. These natural assets are prerequisites for the preservation of a climate favouring well-being of flora and fauna but also well-being of the human society and economy (see Makarieva, Sect. 3.3). Innovative concepts of land management and land use planning are required to satisfy the needs of nature and mankind alike.

The Water Frame Work Directive of the EU may be considered a valuable tool of managing urban areas and the hinterland alike. It is a guideline designed to stabilize the function of ecosystems with the aim to convert and maintain natural water bodies in "good shape" and at "high ecological potential". Thus, the EU Water Framework Directive serves as a tool to overcome the water and food deficiency crises as well as the ecosystem and biodiversity crisis while strengthening the economic and social status of rural areas in relation to the embedded urban areas.

5.1.5 Technical and Political Challenges

It is essential to underline that the complex threats of the four Earth crises depicted in Fig. 5.1 can only be solved in a sustainable, holistic and integral way. Answers will be partially of technical nature but as important are methods which facilitate political decision making in the attempt to overcome negative feedback caused by non-sustainable behaviour. The traditional generic obligations of the State to provide its citizens with favourable living conditions is valid and important but it is insufficient taking into perspective the threats posed by the four Earth crises. The State authorities are responsible for the functioning the anthroposphere but also for the preservation of the entire ecosphere, at least within the boundaries of national jurisdiction. In a time of rapid global change also of complexity (see Dewilde, Sect. 2.1; Mainzer, Sect. 6.1) this task is tremendously encouraging.

As Karl Wittfogel (1957) describes in his book entitled "The Oriental Despotism" those activities require a knowledgeable "bureaucracy" armed with a robust executive power, also called "total power". In ancient times, the rulers of countries in arid areas (Egypt, Mesopotamia, parts of India and China) established such an entrepreneurial administration. The aim was, originally, to provide the basis of a flourishing livelihood. Wittfogel suggests that rise and stability of ancient civilization was tightly linked to the unique power of such administrations which treated people as instruments in the battle against natural catastrophes not allowing any free will. He calls this form of government "hydraulic despotism" in contrast to modern time "democracy" hampered by the all-mighty "not in my backyard" mentality.

Now, that our global civilization is confronted with tremendous challenges caused by the human society itself, and realizing that effective treatment of fresh water resources is a decisive factor of wellbeing of people, stability of life supporting ecological systems, the functioning of economies and the resilience of societies Wittfogel makes us considering to develop a modern form of robust executive hydraulic management based on well-founded democratic principles, capable of organizing our already over-populated planet to the benefit of humankind. This verbal construct—despotisms controlled by democracy—sound contradictorily. History tells us that the aims and goals of modern water management and water technology can only materialize when the administration on duty gets effectively controlled by democratic institutions.

A two-step strategy appears to be taken: Appropriate technical and organisational solutions have to be developed and implemented at the national or transnational level. Moreover it should be realized that the availability of sustainable solutions affects the world marked, assisted by the theoretical knowledge about the pressure and the translation into the national and international policy. In the light of this process it is necessary to increase the dialog between science, engineering and the political decision makers including the governmental regulators.

5.1.6 Concluding Remarks

Sustainable water management plays a crucial role in a world populated by a rapidly growing number of people, by an increasing density of urban areas, and by globalisation of the lifestyle which developed in the industrialized countries over the past 100 years. Health and well-being of people, economic prosperity and stability of societies depend on uninterrupted supply of safe water in sufficient quantity and quality, well elaborated collection and treatment of the used water, and above all on responsible management of the water bodies from which the water is taken, and into which the used water and its components is taken back, hopefully after proper purification.

Water management shall not be envisioned as a stand-alone issue, however. For instance, water management and energy management are inextricably linked. In order to achieve and maintain sustainable development of societies and economies design, installation and regular maintenance of water and energy management systems is a major obligation. Fine-tuned cross-linkage of water and energy management is a matter of national security as Allan Hoffman (2004) clearly stated at a conference held in November 2009 in Copenhagen.

The term "sustainability" stands for human activities kept in the limits of the generic capacity of the Earth's life supporting system—the capacity to deal with resources taken for human consumption in particular. Usage of non-renewable resources such as fossil water and fossil fuel can only be tolerated as a temporary solution. Emphatic and straight forward attempts are to be made at the highest possible speed to develop and implement substituting technology and behavioural changes among the users.

Water is needed for energy generation (hydropower, steam generation, cooling), and energy is needed for water supply (operation of pumps, purification processes, desalination plants and so on). Water and energy are needed by people but also by agriculture to produce food. Enterprises and industries need water and energy to generate goods and services. It is common practice but basically impermissible to manage water and energy supply as a unique economic issue. Sustainable development can set on path only when taking a cross-sectoral approach.

The necessity to further develop and integrate sustainable water and energy management is amplified by a variety of global changes. Major drivers of these changes are the growing number of people on Earth, the steady increase of the migration rate towards urban areas and world-wide adoption of the lifestyle practiced in the industrialized countries. As a consequence, the demand of energy, water, food, medical care, land, commodities and luxury goods is sky-rocking. On the global as well as on the local scale mankind is facing serious crises which are to be solved to keep the local economies, societies and the environment equally in balance. The global warming and climate change crisis is just one of the Earth crises which need to be tackled. As important is the crisis caused by the ecosystem and biodiversity loss, the social and economic crisis and, last but not least, the water and food crisis. It is important to understand that all these crises are interlinked. Sectoral approaches to crisis management are insufficient, therefore. What is needed is a holistic, overarching approach.

Because it is crucial to keep the generic function of the ecology-driven Earth system intact, responsible, cross-sectoral and cross-national water management is highly important but exercising this responsibility is anything but trivial. Decisions on concrete actions must be made not only in view of factors affecting the global and local ecosystems. In addition, a variety of human factors are to be taken into account, growth of the human population for instance, demographic changes, changes of lifestyles, the power of traditions and religious concerns, and many others. Considering the complex nature of the water sector, developing innovative tools to facilitate decision making processes in a world of complexity and heterogeneity is an urgent task.

On a local scale, development of innovative concepts and methods to manage urban areas is required. Here people are facing rapid growth in the demand for water and food, and rapid deterioration of water resources, both quantitatively and qualitatively. Over-fertilization of agricultural land, extensive application of pesticides to mono-agricultures, and sea water intrusion into estuaries and aquifers are among the reasons for the decrease in water quality. To be able to supply people, industry and farmlands with safe water, and to be able to maintain a reasonable public health status, economic prosperity and social stability, use of unconventional sources of water, including wastewater, are to be considered. The cascading use and re-use of water in cities must be seriously considered. Decision makers of political institutions as well as water users must be prepared for a shift from the traditional concept of single-pass water usage towards the concepts of recovery and re-use. The public must understand that wastewater is not a nuisance but a valuable resource.

Water re-use does have risks which should not be underestimated. Keeping water circulating in an urban setting will almost automatically lead to the accumulation of trace-substances. To avoid detrimental health effects and human fertility loss, innovative treatment technology is required to make water re-use a safe, robust and cost-effective method.

Decentralized systems to treat and reuse water require special attention by water authorities responsible for public health. To be developed and implemented are not only high-quality appliances and treatment processes, but also management structures to provide remote control, safety and convenience.

This applies for the urban as well as for the rural areas. The latter are known to contribute substantially to the sustainability of the whole water system. In consequence, rural areas have to be developed with the same whole-heartedness.

Taking all such issues into account, it is obvious that the water sector has not reached a mature state. On the contrary, the R&D demands of the water sector, the need to invent and implement innovative concepts and techniques, and the need to educate operators, regulators and people in general are even greater as global warming, urbanization, lifestyle changes and the desire for comfort are proceeding at an increasing rate. It is time for investment in research, technology development, innovative infrastructure and management methods, a shift in public attitudes. And last but not least, turning away from the well-established but ill-conceived anthropocentric worldview is a must.

5.1.7 Acknowledgement

This article refers to a similar article authored by Wilderer, Grambow and Meng and published in Kauffman J, Lee K-M (eds). Handbook of Sustainable Engineering. Springer, New York (Wilderer et al. 2013).

5.2 Decentralization of Wastewater Treatment and Reuse

Asher Brenner

5.2.1 Key Messages

Decentralization is not a magic solution for sustainable wastewater management.

5.2.2 Introduction

Sustainable wastewater management means meeting both present and future needs for reliable and efficient treatment, and securing effluent disposal or reuse with minimal health risks and ecological damage. However, the ever-increasing urbanization and population density ultimately may cause a long-term concentration buildup of toxic substances in the closed-loop cycle of water supply and wastewater treatment and reuse (Kolpin et al. 2001; Pal et al. 2014). This problem of emerging pathogens and toxic organic constituents requires judicious design of water and wastewater systems.

5.2.3 Is Nature Undefeated?

The "resilience potential" of nature has dramatically changed over the twentieth century. Disposal of huge amounts of organic compounds and nutrients, oil spills, and uncontrolled industrial waste management have caused pollution of rivers, groundwater, and the sea. A remarkable example is the "ecological death" of many rivers all over the world and the failure of their "self-purification" capability to overcome the anthropogenic intervention (see oxygen-sag model simulation, Brenner et al. 2005). Decentralized wastewater treatment and reuse systems should therefore be uniquely based on the aim to have a lower impact on nature than centralized systems.

5.2.4 Decentralized Wastewater Management

Decentralized wastewater management (DWM) has become a new fashion in recent years since it may offer economic, social, and environmental advantages (Fane and Fane 2005). Another motivation for such systems is in areas with limited availability of potable water. Thus, reuse of treated wastewater can supply precious water for non-potable reuse options such as gardening or toilet flushing, and consequently will contribute to water saving and sustainability. Decentralized wastewater treatment systems incorporate collection, treatment, reuse, and disposal of wastewater of various types, including domestic, commercial, or industrial wastes. These systems can be applied for local urban regions, small communities, public institutions, touristic sites, etc. It has been claimed that these small-scale systems can guarantee the same safety and reliability as large-scale centralized systems. However, there can be no compromise on the need to design, operate, and control these systems to meet environmental standards and avoid pollution and health hazards. Sometimes there is no choice other than the application of DWM. This can happen in urban areas where centralized wastewater facilities cannot be expanded, or when there are constructional limitations of collection sewers, or disposal/reuse restrictions. The development of new technologies for wastewater treatment have yielded small and compact systems that can be applied for low wastewater flow rates, suitable for decentralized systems. One example is the membrane bioreactor (MBR) which incorporates secondary biological treatment and tertiary membrane filtration, resulting in high quality effluents.

5.2.5 *Wastewater Treatment Standards for Centralized and Decentralized Systems*

Massive production and use of many anthropogenic chemicals have resulted in the release of toxic and problematic materials into the environment (such as the so called organic micro-pollutants—OMPs, including pharmaceuticals and personal care products—PPCPs, some of which are considered endocrine disrupting compounds—EDCs). Other concerns include newly-emerging pathogens that jeopardize human health. Future standards of effluent quality for reuse or dispersal will probably become more stringent. This means that more-sophisticated technologies will have to be implemented; including advanced oxidation processes, activated carbon adsorption, and tight membrane separation processes such as nanofiltration and reverse osmosis (RO). The formula to define the optimal configuration of these progressive technologies in a quaternary treatment stage has not been developed yet. Furthermore, recent studies indicate that even RO (following MBR treatment) cannot be considered an ultimate treatment method for OMPs (Sahar et al. 2011). Since the new standards and the consequent required technologies must be equal for centralized and decentralized systems, several doubts arise regarding management and control of small decentralized systems. In this regard, Table 5.1 presents common wastewater treatment stages and processes, and outlines their capability and complexity. A compromise between the two opposite management edges may be the semi-centralized approach, which takes into account limitations and advantages of both concepts (Bieker et al. 2010).

5.2.6 *Existence of Sewer Systems as a Key Factor in Decision*

DWM can offer economic, environmental, and social advantages over centralized systems. However, like other water management systems they should be properly designed, operated, and maintained. For isolated communities there is no choice other than DWM. However for other urban options, the size and the location of the community applying DWM might pose technical and environmental limitations. There are also incidents of disposal of untreated sewage to the environment following process failure, or sometimes intentionally. As a consequence, many problematic compounds may find their way into natural water systems. Thus, existence of a sewer system (draining into a central facility) can compensate these limitations and serve as a safety component in failures of DWM systems.

Table 5.1 Efficiency and complexity of wastewater treatment processes

Type of treatment stage or process	Pollutants removed	Complexity of operation and control
Natural biological systems (oxidation ponds, constructed wetlands)	Organic matter	L
Secondary mechanical treatment (activated sludge)	Organic matter, nutrients	M
Tertiary treatment (granular or membrane filtration + disinfection)	Pathogens	M
Quaternary treatment (activated carbon, chemical oxidation, membrane desalination)	Pathogens (including viruses), OMPs	H

L low, *M* medium, *H* high

5.2.7 *Conclusion*

Integration of decentralized and semi-centralized systems in urban areas depends on the projected population growth, layout of sewer system, availability of potable water, and potential reuse applications. There is no single or magic solution for wastewater treatment or reuse. Wastewater treatment processes vary and there is no universal approach. It is possible to achieve any effluent quality standards in large or in small-scale treatment processes. However, proper management, control, and performance standards must be kept for both centralized and decentralized systems. Protection of human health and of the environment is the ultimate goal of sustainable wastewater management.

5.3 Centralized and Dezentralized Structures in Solid Waste Management

Werner P. Bauer

5.3.1 *Key Messages*

To manage the plethora of today's goods and tomorrow's waste we need both, individual initiative and strong environmental governance. In other words, we need central as well as decentral initiatives. As Bill Gates once said: "*In a globalized world with interconnected people the power of [the] crowd will enforce local solutions*". A strong environmental governance must be based on knowledge, transparency and courageous people, persistent at communicating their fieldwork.

5.3.2 Introduction

In its catalogue, a German fashion store offers a special sweater with the words "Initially, you will wear this sweater under your suit at the opera, or at family celebrations or at dinners for 2. You will wear it with a jacket to jeans or chinos. Later on, marked by everyday use, you will still love your sweater and wear it when lounging in front of the TV in the evenings as a top for your pajama trousers. And finally, if you cannot separate from it, your sweater will end his career as a garden outfit when you mow the lawn." There is nothing to worry about when this worn out cloth finally becomes waste material.

All things should be reused and treated properly. But nowadays such a special relationship to things has become quite an exception.

So the first and always most important decentralized issue with waste is, how people handle the things they use and what things they buy. Good quality products definitely have a longer lifecycle. If you buy handcrafted shoes from Budapest, they will serve you for more than 20 years. Buying an expensive mechanical clock will save batteries.

However, manufacturing increasingly cheaper products is a direct result of globalization, population growth, poverty (not only) in developing countries and the buyer's desire a higher standard of living. In addition to these global patterns of consumerist culture, the aspiration to dress and act according to prevalent trends leads to short-term usage of products and an inundation of solid waste formation all over the globe. As consumers base purchasing decisions on their momentary financial situation and not on the possible life expectancy of the products (*their sustainability*), people's purchasing habits and global trade are always directly linked to the production of waste.

"The total merchandise trade grew from $0.3 trillion in 1971 to $18.3 trillion in 2011"[1]—and in the end all goods become waste.

In poor, undeveloped countries even today people carelessly litter the streets with worthless things.

Waste management activities mainly aim at ensuring public health and cleanliness in urban areas, which—when organized village by village—lead to countless dumping sites. Already in the 1970s, Bavaria with approximately 10 million inhabitants had over 6000 dumpsites close to rural communities and even in the cities.

As a first step to optimize waste management, many countries closed these local disposal sites and designed sanitary landfills, which comprised the waste of larger areas (more than 100,000 inhabitants per landfill). This step was necessary to prevent polluting the groundwater with leachate from unsealed sites and to reduce or prevent the emission of greenhouse gases from open deposits.

[1] Mavropoulos et al. (2012).

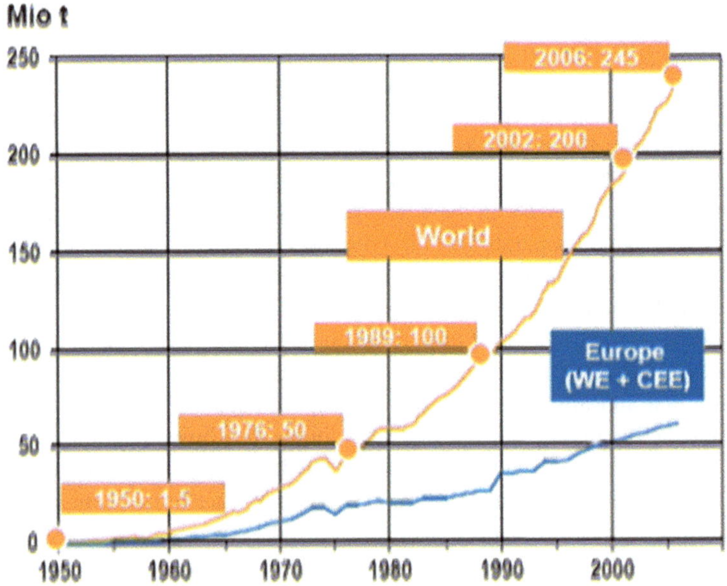

Fig. 5.5 Increase of plastic production from 1950 to 2006 (Ralf Strobel (2009) "Nach uns die Plastikflut", Die ZEIT online, 18th December 2009)

In Bavaria today, these measures have lowered the number of places where industry and municipalities can deposit their waste to less than 50 landfills. Deposits of MSW (>5 % organic) in Germany stopped in July 2005.

In terms of federal state planning, the demand for modern waste management structures could serve as a central guideline to mitigate the effects of climate change, but very often it is also necessary from a decentralized point of view to protect the local drinking water.

To centralize undefined waste depositing in an organized large-scale landfill requires being able to discern different kinds of waste and to manage these diverse mass flows at different places.

Waste in the 1950ies was a totally different matter compared to today's waste. Since the introduction of synthetic pulps in the 1940ties the world's production of plastic materials has increased over the past 20 years up to "265 million tons in 2010"[2] (Fig. 5.5)

Today's waste is linked to the production of plastics, chemical products and to the fact that industrial waste might have toxic substances.

[2] IETC, International Environmental Technology Centre, PROJECT CONVERTING WASTE PLASTIC INTO FUEL.

5.3.3 Landfill

With this in mind, a modern landfill will separate areas for municipal waste from those for toxic waste. And they will take into account that inert demolition material has to be separated from organic and toxic fractions of demolition waste.

The best way to control these mass flows is to establish a tipping fee, which is calculated based on the fact that a waste disposal site has to be operated for more than 50 years after stopping the disposal.

If users of the landfill only pay a small fee, calculated to amortize the expenses and short-term operational costs, landfilling is the cheapest way to handle waste and recycling waste will thus remain a task of the waste-pickers.

Today, however, we are aware that leachate and gases from a landfill cause environmental problems for many years and have to be retained and treated during the whole aftercare period.

To clarify: tipping fees for a landfill are calculated based on

- Costs for the purchase of real estate
- External preparation of land and compensation areas
- Construction period
  - Surface conditioning
  - Sealing (mineral sealing, plastic liner, protective membrane, geotextile)
  - Leachate pipeline
- Internal preparation of land, traffic areas
- Leachate treatment facilities, basins, etc.
- Premises and weight bridges
- Degasification
  - Gas recovery
  - Gas wells
  - Gas control systems
- Surface sealing, re-cultivation
- Reloading station
- Detailed planning and building inspection
- Operating period with the operational costs per year
  - Personnel
  - Energy
  - Degasification (includes costs for maintenance of the gas collection as well as the gas recovery (gas motors and compressors) including wear material and external service
  - Leachate treatment and cleaning
  - Vehicle fleet
  - Administration

- Aftercare period (30–50 years)
  - Personnel
  - Energy
  - Degasification
  - Leachate treatment and cleaning
  - Administration
- Renaturalisation
- Insurances
  - Environmental liability Insurance
  - Directors' and officers' professional liability insurance
  - Accident insurance for employees

So, to calculate a sufficient tipping fee might be a decision to make on a decentralized basis; a decision for those responsible for local waste management at a regional level. An open discussion with the local population will help those who oppose and will diminish corruption.

Well-calculated tipping fees are the best foundation for the next steps of waste management. However, according to the ISWA Waste Management 2030+ Report[3] "...thinking globally, the massive development of new sanitary landfills is the only realistic and achievable option for a universal step forward ...." we need to ensure that landfills are well-designed and well operated. His arguments point to the necessary and expensive infrastructure of waste management for MSW, that "is delivered much slower than the rapid growth of waste generation".

Besides this there is no room for new landfills.

In large cities the next inevitable steps in waste management development are to install incineration plants or even better waste-to-energy plants.

A comparison of landfill and incineration according to the method of Total Costs of Ownership (TCO), as regards total lifecycle and including the aftercare of landfills show that with a price of Euro Cents 2.5 per kWh of steam use, the waste fees of the waste-to-energy plant scenario lie below the waste fees of the landfill scenario (Fig. 5.6).

5.3.4 Incineration

Waste-to-energy plants have to be situated close to industry or in the middle of the urban areas, where people require the majority of the energy produced and where it is possible to lower transportation costs.

[3] Mavropoulos (2014).

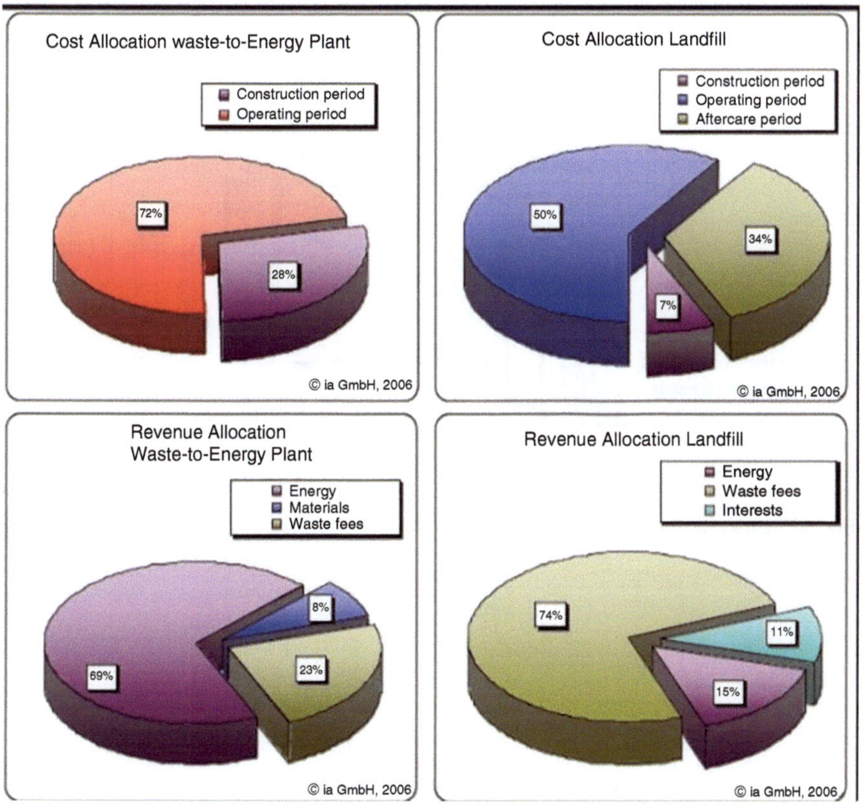

Fig. 5.6 Cost and revenue allocations waste-to-energy plant—landfill CEWEP—lifecycle comparison of waste-to-energy plant and landfill by the method of total costs of ownership (TCO), Charts: ia GmbH

Besides greater cities, also densely populated areas usually have waste-to-energy plants (areas of approximately over 500,000 inhabitants) (Fig. 5.7).

Professional recycling in a highly effective circular flow is the next step—but let us first review the basics of dumping.

On uncontrolled dumps, waste is burnt almost daily. This means that mostly, waste is incompletely combusted and the process is totally different to that in a waste-to-energy plant (Fig. 5.8).

As the picture above shows, moving from landfill to the system of incineration is only one possibility to upgrade a local waste management system. It is necessary to reduce emissions and optimize the process in all structures—both in landfill systems and in incineration systems. To seal the surface and to install gas utilization for the landfill—which could be organized decentrally—is just as necessary as taking a step towards greater centralization, which is needed for the implementation of an incineration plant.

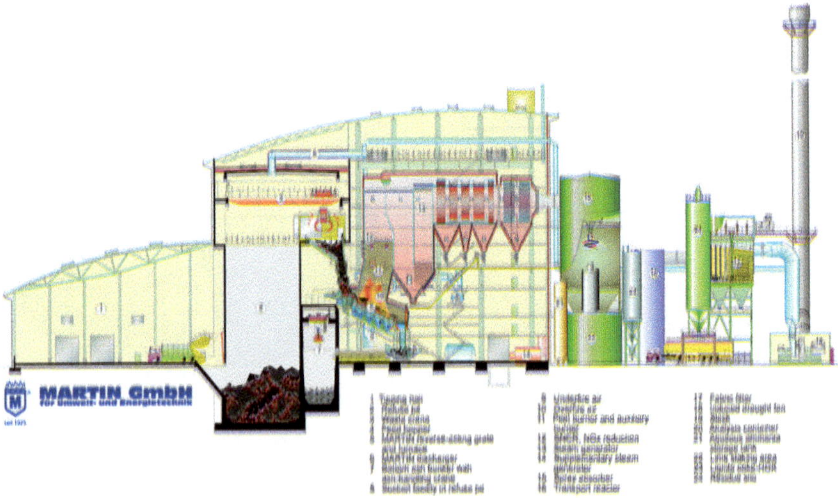

Fig. 5.7 Cross section waste-to-energy plant

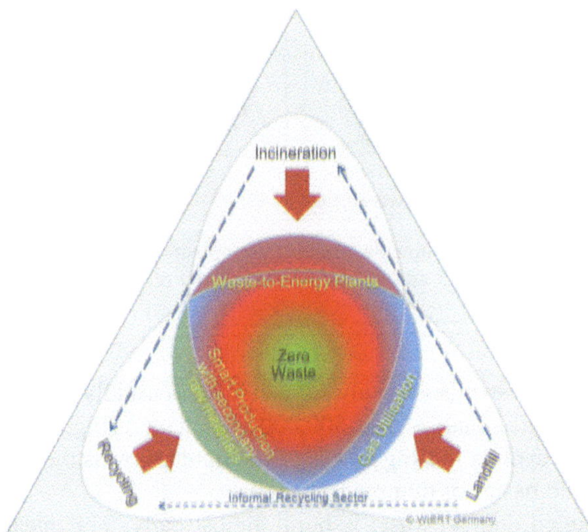

Fig. 5.8 Waste-to-energy an evolution from landfill, incineration and recycling to zero waste, Chart: WtERT Germany GmbH

Higher quality of solutions for SWM, both in landfill solutions and in incineration lead to higher costs and fees. Recycling processes too can be optimized and made more efficient.

However, even if high costs of incineration plants boost professional recycling activities the most, there are some other drivers to "reduce—reuse—recycle".

5.3.5 Drivers of Recycling

Recycling is a process that brings back waste materials into the production of new products. The idea of recycling is that using secondary raw material reduces consumption, energy, pollution and costs.

For years, used glass, old paper and scrap metal have been procured as cheap, secondary raw materials to create new products.

With this in mind, millions of people in the third world live on dumpsites and pick useable waste for their livelihood. In this structure called the "informal recycling sector" the driver is poverty. But beside this "in many developing countries, the informal recycling sector achieves notable recycling rates.[4]" (Fig. 5.9)

Driven by poverty this method of waste management is at an extremely low level of decentralisation.

Other crucial motivations to recycle are

- Conservation of resources and energy and
- Protection of water and soil.

Considering the relation between waste and water management, phosphorous as a very restricted natural resource, will have a key role in the latest and future handling strategies. This applies to sludge from wastewater cleaning as well as from MSW.

Fig. 5.9 The Informal recycling sector, "Recicladores" on a dumpsite in Colombia, Pictures Werner P. Bauer

[4] Mavropoulos et al. (2012).

In January 2015, 11 main German associations elaborated a strategy to deal with sludge consensually. Essential receivables are:

Only good quality sludge should be used for agriculture. When recycling, phosphorous, nitrogen and other organic substances were used highly efficiently. In future, loaded sludge will be burned only in incineration plants focused on burning sludge so that phosphorous can be retrieved from the sludge-ash. The outcome of the recovery should be an effective and useful product like fertilizer for example.

Even keeping phosphorous as a separate organic waste in a selective collection is a special aim of modern waste management. Fermentation plants are able to use most of the energy of the waste by producing bio-methane or natural biogas and digestates, which makes phosphorous available for farming. There are fermentation plants at a very small, decentralized level and as one recycling plant per approx. 500,000 inhabitants. Linked to industry or other users of energy this central fermentation plant serves also as a waste-to-energy plant, an incineration facility with a good usage of steam.

According to a study,[5] "the biggest cities in Brazil generate about 18.9 million tons of municipal solid waste per year (2011), of which 51.5 % is biogenic matter." Burning provides energy but using this in fermentation plants and burning the other 48.5 % provides energy and organic waste including phosphorous.

5.3.6 *Drivers of Reduction*

A now common interest to mitigate marine littering is another specific driver of recycling or as this example shows, a driver of reducing waste, which seems to be possible only at a central level.

Since January 1st 2011, Italy[6] has banned non-biodegradable plastic bags by law. Italy was one of Europe's top consumers of plastic bags: more than 330 plastic bags per person per year. After introducing this law, stores may only offer biodegradable, cloth or paper bags.

When classifying this activity as a centralized action, it is necessary to realize that in order to stimulate the legislation of a whole nation sometimes only a few opinion leaders are necessary, who continuously communicate these topics to the people.

In a German newspaper,[7] Bill Gates, one of the most highly influential persons in the world, was quoted to say that voters should put their government under pressure. With global problems the power of the people is more important than a good deed of one single person.

[5] de Souza et al. (2013).

[6] http://www.plasticsnews.com/article/20110110/NEWS/301109941/italy-imposes-pe-bag-ban

[7] Kreye (2014).

5.3.7 From "Reduce: Reuse—Recycling" to "Zero Waste"

Besides Jack Johnson, the American songwriter who wrote his 3R-song, the currently strongest driver of the idea of "reduce—reuse—recycling" is the European Commission:

The directive on waste of the EU[8] has established a legal framework for the treatment of waste within the EU. It aims at protecting the environment and human health through the prevention of the harmful effects of waste generation and waste management.

Article 4 shows the following waste hierarchy, which is listed in order of priority:

- Prevention
- Preparing for reuse
- Recycling
- Other recovery, notably energy recovery
- Disposal

"Member States can implement legislative measures with a view to reinforcing this waste treatment hierarchy. However, they should ensure that waste management does not endanger human health and is not harmful to the environment.[9]"

In anticipation of the revision of the European Waste Framework Directive, the commissioner of environmental matters, Janez Potocnik, declared in his opening speech for the "Green Week" in June 2014 in Brussels that: "In a circular economy there can be no place for waste".

Linda Wagner[10] said at a symposium in July 2014 that the EU Commission linked the management of waste and resources to the economy: "Circular economy—saving resources, creating jobs"

As half of MSW in Europe is still deposited in dumps or landfills there is still a lot to do and it is very noteworthy that the arguments are of an economical and not only an ecological nature.

"Current European waste policy does not mainly aim to treat waste streams but rather place in the foreground of interest the complete supply chain of a product. Waste prevention and re-use do have the highest priority and they take effect before the end-of-life phase of a product or a material is reached. Recycling only takes the third place whereas recovery and disposal represent the least favourable options.[11]"

So far "reduce—reuse—recycle" aims at zero waste.

"Usually the term "Zero Waste" is associated with the perspective of a complete circular economy that is therefore referred to as an illusion. This viewpoint of waste management stakeholders is quite understandable since the political demand for an increased effort on the side of waste Management players towards a full circular

[8] EU, 2008, Official Journal of the European Union of 22.11.2008.

[9] EU, 2008, Official Journal of the European Union of 22.11.2008.

[10] Wagner (2014).

[11] Bartl (2014).

economy with closed material loops from a position starting at the end of our economic system is limited regarding its success. Under the given circumstances especially the waste related tasks of removal and concentration regarding the included contaminants are pointed out. "Zero Waste" will succeed when every waste will make sense in a new state.[12]"

However, from a global point of view we must admit that the major trends for the next decades might lead to a totally different situation.

5.3.8 *Waste Management 2030+*

In his Waste Management 2030+ Report, Antonis Mavropoulos points out

- Growing waste volumes because of the population growth and "its consumption patterns which are controlled by the evolution of Gross Domestic Product per Capita (GDP/c)[13]".
- Changing of waste composition
  - The vast demand for agricultural goods that will lead to more organic waste,
  - The production of more and more complex products will bring a rapidly growing stream of electronic waste (e-waste) and
  - A stream of nanomaterials, that are slowly but steadily becoming waste.

Antonis Mavropoulos, CEO of ISWA, considers that the European Union as an example of more or less the most advanced continental waste management system. But he also "wonders how successful the EU waste management would be without China and Asia to absorb both legal and illegal waste shipments?" Traffiking waste on a boat to China is much cheaper than recycling it in good old Europe.

Exporting waste illegally to poor countries has become a vast and growing international business, we have to face the increasing volumes and rapidly changing composition of waste.

References

References of Section 5.1

Agenda 21: Earth Summit – The United Nations Programme of Action from Rio (1993) United Nation. ISBN 13: 9789211005097

Anonymous (2000a) Integrated Water Resources Management. Global Water Partnership, Stockholm, Sweden

---

[12] R. Mittermayr, from a open discussion on the topic "What is Zero Waste" in www.wtert.eu

[13] Mavropoulos (2014).

Anonymous (2000b). Directive 2000/60/EC of the European Parliament and of the Council establishing a framework for the Community action in the field of water policy. Official Journal of the EU, OJ L 327
Anonymous (2008) Zugspitze declaration: dealing wisely with the planet earth. http://www.euro-acad.eu/downloads/memorandas/zugspitze_declaration-final.pdf
Cornel P (2007) Semi-centralized supply and treatment systems for fast growing areas. In: Huber HH, Wilderer PA, Paris S (eds) Water supply and sanitation for all. Hans Huber AG, Berching, pp 211–212. ISBN 13978-1-84339-514-0
Grambow M (2008), Wassermanagement, Integriertes Wasserresourcen management (IWRM) von der Vision zur praktischen Umsetzung (Water management, integrated water resource management (IWRM) from vision to practical implementation), Vieweg
Grambow M (ed) (2013) Nachhaltige Wasserbewirtschaftung (Sustsainable water management). Springer Vieweg, Wiesbaden, Germany
Hoffman A (2004) Energy Secur J, IAGS, 13 Aug 2004
Kauffman J, Bailkey M (2000) Farming inside cities: entrepreneurial urban agriculture in the United States. Chicago, Lincoln Institute of Land Policy. Working paper
Larsen TA, Gujer W (2000) Waste design and source control lead to flexibility in wastewater management. IWA World Water Congress, Paris, pp 93–300, book 5
Rodell M, Velicogna J, Famiglietti JS (2009) Grace reveals massive depletion of groundwater in NW India. Nature 460:699–1002
Taylor R (2014) When wells run dry. Nature 516:179–180
Tortajada C (2006) Water management in Singapore. Water Resour Dev 22:227–240
Wilderer PA (2004) Applying sustainable water management concepts in rural and urban areas: some thoughts about reasons, means and needs. Water Sci Technol 49(7):7–16
Wilderer PA (2005) UN water action decade: a unique challenge and chance for water engineers. Water Sci Technol 51(8):99–107
Wilderer PA (2007) Sustainable water resources management: the science behind the scene. J Sust Sci 2:1–4
Wilderer PA (2009) Climate change and its consequences for water and wastewater management. J Front Environ Sci 3–1:1–11
Wilderer PA (2014) Stadtnahe Regionen intelligent entwickeln (Intelligent development of urban regions) Urban 2.0, pp 38–39
Wilderer PA, Grambow M, Meng W (2013) Sustainable water management in response to global changes. In: Kauffman J, Lee K-M (eds) Handbook of sustainable engineering. Springer, New York
Wilson G (2005). Singapore's new business opportunity: food from the roof. Retrieved from http://www.greenroofs.com/archives/gf_nov-dec05.htm
Wittfogel KA (1957) Oriental despotism: a comparative study of total power. Yale University Press, New Haven

References of Section 5.2

Bieker S, Cornel P, Wagner M (2010) Semi-centralized supply and treatment systems: integrated infrastructure solutions for fast growing urban areas. Water Sci Technol 61(11):2905–2913
Brenner A, Shacham M, Cutlip MB (2005) Applications of mathematical software packages for modelling and simulations in environmental engineering education. Environ Model Softw 20 (10):1307–1313
Fane AG, Fane SA (2005) The role of membrane technology in sustainable decentralized wastewater systems. Water Sci Technol 51(10):317–325

Kolpin DW, Furlong ET, Meyer MT, Thurman EM, Zaugg SD, Barber LB, Buxton HB (2001) Pharmaceuticals, hormones, and other organic wastewater substances in U.S. streams, 1999-2000: a national reconnaissance. Environ Sci Technol 36(6):1202–1211

Pal A, He Y, Jekel M, Reinhard M, Yew-Hoong GK (2014) Emerging contaminants of public health significance as water quality indicator compounds in the urban water cycle. Environ Int 71:46–62

Sahar E, David I, Gelman Y, Cikurel H, Aharoni A, Messalem R, Brenner A (2011) The use of RO to remove emerging micropollutants following CAS/UF or MBR treatment of municipal wastewater. Desalination 273:142–147

References of Section 5.3

Bartl A (2014) Moving from recycling to waste prevention: a review of barriers and enables (ISWA article)

Bauer WP et al (2006) Total costs of ownership: comparison incineration plant-landfill. Ordered by the Bavarian State Ministry of Environment within the scope of the Technical Cooperation Project Bavaria Sao Paulo in February 2006. Waste-to-Energy Research and Technology Council, Munich

de Souza SNM et al (2013) Technical potential of electricity production from municipal solid waste disposed in the biggest cities in Brazil: landfill gas, biogas and thermal treatment. ISWA

Kreye A (2014) Bloß kein Mitleid. Süddeutsche Zeitung of 20–21 Dec 2014, p 49

Mavropoulos A (2014) ISWA waste management 2030 + the future of waste management in an overcrowded planet. Waste Management World, Tulsa, OK

Mavropoulos A et al (2012) Globalisation and waste management. Final report by the ISWA Task Force

Strobel R (2009) Nach uns die Plastikflut. Die ZEIT Online, 18 Dec 2009

Wagner L (2014) Höhere Recyclingquoten und Deponierungsverbote in einer neuen Kreislaufwirtschaft in Europa in Kommunen in der Verantwortung. Tagung der ia GmbH 2014

Chapter 6
Communication, Mobility and Logistics

Klaus Mainzer, Verena Huber, and Joachim Schütter

6.1 Challenges of Complexity in the Age of Globalization and Big Data

Klaus Mainzer

6.1.1 Key Messages

Self-organization in nature and society favors decentralization, but nevertheless needs control functions of governance.

6.1.2 Overview

The global problems of mankind (e.g., energy, mobility, urbanization, nutrition, financial markets, communication) are cross-over to specialized disciplines of science and need interdisciplinary studies in systems science (1). In sociotechnical systems, information and communication technology (ICT) is growing together

K. Mainzer (✉); Corresponding author for Section 6.1
Technical University of Munich, Munich, Germany
e-mail: mainzer@tum.de

V. Huber (✉); Corresponding author for Section 6.2
Technical University of Munich, Munich, Germany
e-mail: verena.huber@tum.de

J. Schütter
Bavarian Institute for Environment, Augsburg, Germany
e-mail: joachim.schuetter@lfu.bayern.de

P.A. Wilderer, M. Grambow (eds.), *Global Stability through Decentralization?*,
Strategies for Sustainability, DOI 10.1007/978-3-319-24358-0_6

with societal infrastructures (e.g., smart grids, smart cities), in order to handle the complexity of human civilization (2). In the age of big data, information and communication technologies (ICT) promise support of knowledge sharing through global online participation. But big data technologies only deliver technical support, no competence of interdisciplinary problem solving. Increasing complexity of our civilization needs reflection on the foundations and laws of systems dynamics in order to support global governance (3).

6.1.3 Globalization and Systems Dynamics

In the age of globalization, the Earth system grows together with human civilization. Climate and ecological systems can no longer be separated from human civilization, but depend on industrial growth and energy policies (Fig. 6.1). Global communication networks and infrastructures as well as financial dependencies of banks and states are driven by nonlinear dynamics of complex systems. One of the main insights of nonlinear dynamics is ("decentralized") self-organization and the emergence of systemic risks which are caused by the interactions of many factors and players in the whole system of Earth. The emergence of systemic risks from complex systems dynamics is a challenge for control tasks in engineering sciences as well as regulation and governance in social systems. We need modeling in systems science with early warning systems in the technical and natural sciences as well as economics and politics. In higher education, the awareness of global networks must be supported by interdisciplinary courses of systems science.

Fig. 6.1 Nonlinear dynamics of earth system and human civilization

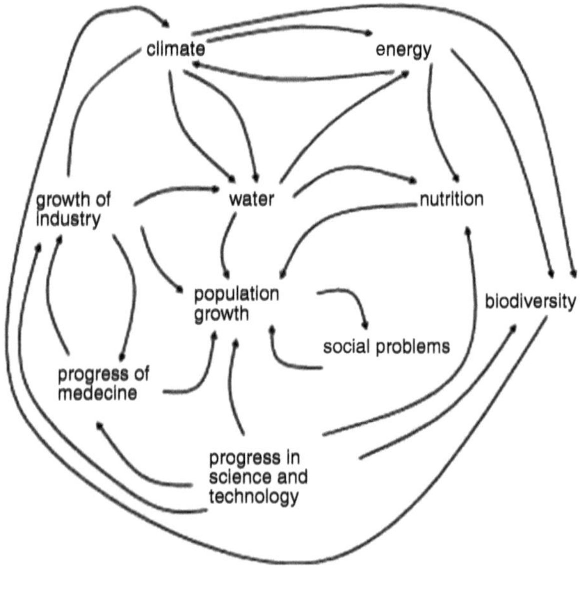

Systems science offers a common language of natural, engineering, economic, and social sciences which is necessary to communicate with different disciplinary backgrounds of education.

Interdisciplinary courses of systems science must start with basic concepts of modeling: Models are formal descriptions of systems in different sciences. They refer in particular to natural systems in astronomy (e.g., planetary systems), physics (e.g., atomic systems), chemistry (e.g., molecular bonds), and biology (e.g., cellular networks), but also to social systems in economics (e.g., financial markets), sociology (e.g., social networks) and political science (e.g., administrative organizations). When engineers analyze a technical system to be controlled or optimized, they also use a mathematical model. In mathematical analysis, engineers can build a model of the system as a hypothesis of how the system should work, or try to estimate how an unforeseeable event could affect the system. Examples are extreme events and risks emerging in complex systems. Similarly, in control of a system, engineers can try out different control approaches in simulations. Simulations are often represented by computer programs and tested on computers. In the natural sciences, the validity of models is tested by derived explanations or predictions which are confirmed or falsified by observations, measurements and experiments. A hypothetical model is a more or less appropriate mapping of reality.

A *mathematical model* usually describes a system by a set of variables and a set of equations that establish relationships between the variables (Gershenfeld 1998; Mainzer 2007a; Yang 2008). A *dynamical system* is characterized by its elements and the time-depending development of their states. The *states* can refer to moving planets, molecules in a gas, gene expressions of proteins in cells, excitation of neurons in a neural net, nutrition of populations in an ecological system, or products in a market system. The *dynamics* of a system, i.e. the change of system states depending on time, can mathematically be described by, e.g., time-depending differential equations. In a more intuitive way, a conservative system is "closed" with respect to external influences and only determined by its intrinsic dynamics. A dissipative system can be considered to be "open" to external influences, e.g., air or other material friction forces. Models of conservative and dissipative systems can also be applied in ecology and economics. In higher education, it is necessary to illustrate basic concepts of systems science by intuitive examples of application:

6.1.3.1 Case Study: Conservative and Dissipative Systems in Ecology

At the beginning of the twentieth century, fishermen in the Adriatic Sea observed a periodic change of numbers in fish populations. These oscillations are caused by the interaction between predator and prey fish. If the predators eat too many prey fish, the number of prey fish and then the number of predators decreases. The result is that the number of prey fish increases, which then leads to an increase in the number of predators. Thus, a cyclic change of both populations occurs. In 1925, the Italian mathematicians Lotka (1956) and Volterra suggested a dynamical model to describe the prey and predator system. Each *state* of the model is determined by the numbers of prey fish and the number of predator fish. So the *state space* of the model is represented by a two-dimensional Euclidean plane with a coordinate for prey fish and a coordinate for predator fish. The observations, over time, of the two populations describe a dotted line in the plane. Births and deaths change the coordinates by integers, a few at a time. To apply continuous dynamics, the dotted lines must be idealized into continuous curves. Obviously, the *Lotka-Volterra model* is closed to other external influences of, e.g., temperature or pollution of the sea. If these external forces of "ecological friction" were added to the model, its dynamics would change the cyclic behaviour.

6.1.3.2 Case Study: Conservative and Dissipative Systems in Economy

In 1967, the economist Goodwin proposed a conservative dynamical model to make the nineteenth-century idea of class struggle in a society mathematically precise (cf. Goodwin 1990; Mainzer 2007b). He considered an economy consisting of workers and capitalists. Workers spend all their income on consumption, while capitalists save all their income. Goodwin used a somewhat modified predator–prey model of Lotka and Volterra. This *conservative model* supports the idea that a capitalist economy is permanently oscillating. Obviously it is superficial, because it does not refer directly to the functional income shares of capitalists and workers or to their population size. But it is mainly its conservative character that makes Goodwin's model seem economically unrealistic. Thus, the model has been made more realistic by the assumption of "*economic friction*". In reality, an economic system cannot be considered as isolated from other dynamical systems. An economic model of coupled oscillatory systems is provided by international trade. In other cases, economic systems are influenced by political interventions. We will come back to these examples later on.

The Lotka-Volterra equations are a simple, but still nonlinear formal system which is fine for educational tasks. Students learn to model the interaction of prey and predators in zoology as well as economics. Modeling in this way is a top down procedure from mathematical equations to applications by appropriate interpretations of variables. In a bottom up approach, we start with a sequence of measurements and ask what the data themselves can tell us about the laws of dynamics. Sequences of data are called *times series*. Time series analysis is used to find types of appropriate equations fitting the data, or to compare the predictions of mathematical models to measurements made in the field of research.

In an ideal case, *time-series analysis* delivers a computer program providing a mathematical model fitting the measured data. But these data-generated models have a severe shortcoming, because they work without any understanding of the physical system. In practice, model building is combined with times-series analysis. Model building is based on knowledge of a physical system, while time-series analysis can be used to detect features of a system, inspiring model building.

For students, it is often inspiring and motivating to learn more about the historical context of scientific discoveries and developments. During the centuries of classical physics, the universe was considered a deterministic and conservative system. A system is said to be *deterministic* when future events are causally set by past events. The astronomer and mathematician P.S. Laplace assumed the total computability and predictability of nature if all natural laws and initial states of celestial bodies are well known. The Laplacean spirit expressed the belief of philosophers in determinism and computability of the world during the eighteenth and nineteenth century. In this historical period, mechanics was a universal paradigm of research. Mechanical machines dominated the first period of industrialization. Laplace was at least right about linear and conservative dynamical systems. In general, a *linear relation* means that the rate of change in a system is proportional to its cause: Small changes cause small effects while large changes cause large effects.

At the end of the nineteenth century, H. Poincaré discovered that celestial mechanics is not a completely computable clockwork, even if it is considered a deterministic and conservative system. The mutual gravitational interactions of more than two celestial bodies (‘*Many-bodies-problem*’) can be illustrated by causal feedback loops analytically represented by nonlinear equations which are sensible with respect to tiny perturbations. Causes and effects are no longer proportional: Tiny deviations in digits behind the decimal point of measurement data may lead to completely different forecasts. This is the reason why attempts to forecast weather fail in an unstable and chaotic situation. In principle, the wing of a butterfly may cause a global change of development (“*butterfly effect*”). The butterfly effect is an immensely important insight for students, in order to understand the nonlinear dynamics in nature, economy, and society.

Typical phenomena of our world, such as weather, climate, the economy and daily life, are much too complex for a simple deterministic description to exist. Even if there is no doubt about the deterministic evolution of, e.g., the atmosphere, the current state whose knowledge would be needed for a deterministic prediction contains too many variables in order to be measurable with sufficient accuracy. Hence, our knowledge does not usually suffice for a deterministic model. Now, in higher education, statistics and probability theory come in. They also deliver basic knowledge for interdisciplinary modeling in systems science.

Actually, very often a stochastic approach is more situated. Ignoring the unobservable details of a complex system, we accept a *lack of knowledge* (Weidlich 2002). Depending on the unobserved details, the observable part may evolve in different ways. However, if we assume a given probability distribution for the unobserved details, then the different evolutions of the observables also appear with specific probabilities. Thus, the lack of knowledge about the system prevents us from deterministic predictions, but allows us to assign probabilities to the different possible future states. It is the task of a time series analysis to extract the necessary information from past data. Again, in higher education, we should support interdisciplinary studies by intuitive examples.

6.1.3.2.1 Example: Power Laws and Risks

In the simplest case of statistical distribution functions, a *Gaussian distribution* has the well-known shape of a clock with exponential tails situated symmetrically to the far left and right of the peak value. Extreme events (e.g., disasters, tsunamis, pandemics, worst case of nuclear power plants) occur in the tails of the probability distributions (Embrechts et al. 2003). Contrary to the Gaussian distribution, probabilistic functions $p(x)$ of heavy tails with extreme fluctuations are mathematically characterized by *power laws*, e.g., $p(x) \sim x^{-\alpha}$ with $\alpha > 0$. Power laws possess scale invariance corresponding to the (at least statistical) self-similarity of their time series of data. Mathematically, this property can be expressed as $p(bx) = b^{-\alpha}p(x)$ meaning that the change of variable x to bx results in a scaling factor independent of x while the shape of distribution p is conserved. So, power laws represent *scale-free* complex systems. The Gutenberg-Richter size distribution of earthquakes is a typical example of natural sciences. Historically, Pareto's distribution law of wealth was the first power law in the social sciences with a fraction of people presumably several times wealthier than the mass of a nation (Mainzer 2007b).

An important part of the modeling process is the *evaluation of an acquired model*. How do we know whether a mathematical model describes the system well? This is not an easy question to answer. We must become aware of these methodological needs and failures. Usually the engineer has a set of measurements from the system which are used in creating the model. Then, if the model was built well, the model will adequately show the relations between system variables for the measurements at hand. The question then becomes: How do we know that the measurement data is a representative set of possible values? Does the model describe well the properties of the system between the measurement data (interpolation)? Does the model describe well events outside the measurement data (extrapolation)?

The mathematical rigor and numerical precision of risk management and asset pricing tools in economy has a tendency to conceal the weakness of models and their assumptions to those who have not developed them and do not know the potential weakness of the assumptions. Even practitioners in economy, finance, or even medicine are often not aware of the conditions and limits of their applied models. In the last financial crisis of 2008, the failing of so-called experts were obvious.

Models are only approximations to the real world dynamics and partially built upon *idealized assumptions*. A typical example is the belief in normal distribution of asset price changes completely neglecting the importance of extreme events. Considerable progress has been made by moving to more sensitive models with *fat-tailed Lévy processes* (Mandelbrot and Hudson 2004; Mainzer and Chua 2013). Of course, such models better capture the intrinsic volatility of markets. But they might again contribute to enhancing the control illusion of the naïve user.

Therefore, market participants and regulators have to become more sensitive towards the potential weakness of risk management models. Since there is not only one true model, robustness should be a key concern. *Model uncertainty* should be taken into account by applying more than a single model. For example, one could rely on probabilistic procedures that cover a whole class of specific models. The theory of robust control provides a toolbox of techniques that could be applied for this purpose.

In macroeconomics, data mining is often driven by the pre-analytic belief in the validity of certain models which should justify political or ideological opinions. The political belief in deregulation of the 1990 years is a typical example. Rather than misusing statistics as a means to illustrate these beliefs, the goal should be to put theoretical models to scientific tests like in the natural sciences.

A chain of specification tests and estimated statistical models for simultaneous systems would provide a benchmark for the tests of models based on economic behavior. Significant and robust relations within a simultaneous system would provide empirical regularities that one would attempt to explain, while the quality of fit of the statistical benchmark would offer a confidence for more ambitious models. Models that do not reproduce (even) approximately the quality of the fit of statistical models would have to be rejected. This methodological criterion also has an aspect of ethical responsibility of researchers: Economic policy models should be theoretically and empirically sound. Economists should avoid giving policy recommendations on the base of models with a weak empirical grounding and should, to the extent possible, make clear to the public how strong the support of the data is for their models and the conclusions drawn from them.

A neglected area of methodology is the degree of connectivity and its interplay with the stability of the complex system. For supervision, one must learn to analyze the network aspects of the financial system, collect appropriate data, define measures of connectivity and perform macro stress testing at the system level. In this way, new measures of financial fragility would be obtained. This would also require a new area of accompanying research and education that looks at agent-based models of the financial system, performs scenario analyses and develops aggregate risk measures. Network theory and the theory of self-organized criticality of highly connected systems would be appropriate starting points (Scheinkman and Woodford 1994; Mainzer and Chua 2013).

Such scientific analysis must be supported by more practical consequences. The hedge fund market is still widely *unregulated*. The interplay between *connectivity*, *leverage* and *system risks* needs to be investigated at the whole level. It is highly likely that extreme leverage levels of interconnected institutions impose dangerous social risks on the public.

On the macroeconomic level, it would be desirable to develop *early warning schemes* that indicate the formation of bubbles. Combinations of indicators with time series techniques could be helpful in detecting deviations of financial or other prices from their long-run averages. Indication of structural change would be a sign of changes of the behavior of market participants of a bubble-type nature (McCauley 2004).

Obviously, there is no single causal model as definitive mapping of reality. But that does not mean a complete deny of mathematical tools and models. We have to consider whole *classes of possible stochastic models* with different weights. They must be combined with a *data-driven methodology* and insights in the factual human behavior and its diversity. Therefore, *psychological* and *sociological case studies* of human behavior under risk conditions (e.g., stakeholders at stock markets, pilots in aircrafts, surgeons in risk surgeries) are necessary. For students of economics, trained in the traditional doctrines and dogmas of economic rationality, it is important to become aware of *bounded rationality*. In experimental economics, decision behavior is already simulated under laboratorial conditions. Even *philosophical ethics* can no longer only argue with arm-chaired considerations and a priori principles, but must relate to empirical observations of factual decision behavior. That is done in the new approaches of experimental ethics. We argue for this kind of *interdisciplinary methodology* which opens new avenues for mathematical modeling in science. In this case, robust stochastic tools are useful, because they are used under restricted conditions and with sensibility for the permanent model ambiguity.

6.1.4 Information and Communication Technology (ITC) and Sociotechnical Systems

Thus, the complex socio-technical challenges of human civilization can no longer be handled without interdisciplinary education of engineering sciences with social sciences and humanities. Examples are efficient concepts of traffic and mobility, complex and intelligent energy networks (e.g., smart grids), urban infrastructures and megacities which are sometimes called cyber-physical systems. The success of large scale infrastructure technologies sensitively depends on the societal, political, and social framework. Therefore, we need integrated research and education of engineering and social sciences with humanities. Integrated research and education lead to sustainable innovation.

6.1.5 Reminder of the Internet as Complex Dynamical System

In a technical co-evolution, global information and communication networks are emerging with surprising similarity to self-organizing ("*decentralized*") neural networks of the human brain. The increasing complexity of the World Wide Web (www) needs intelligent strategies of information retrieval and learning algorithms simulating the synaptic plasticity of a brain (Berners-Lee 1999). The Internet links computers and other telecommunication devices. At the router level, the nodes are the routers, and the edges are their physical connections. At the interdomain level, each domain of hundreds of routers is represented by a single node with at least one route as connection with other nodes. At both levels, the degree distribution follows *a power law of scale-free network* which can be compared with cellular networks in biology. Measurements of the clustering coefficient deliver values differing from random networks and significant clusters. The average paths at the domain level and the router level indicate the small-world property.

6.1.6 Smart Grids as Complex Dynamical Systems

Global information networks are growing together with societal infrastructure. Current examples are complex smart grids of energy. Many energy providers of central generators and decentralized renewable energy resources lead to power delivery networks with increasing complexity. Smart grids mean the integration of the power delivery infrastructure with a unified communication and control network, in order to provide the right information to the right entity at the right time to take the right action. It is a complex information, supply and delivery system, minimizing losses, self-healing and self-organizing (http://ec.europa.eu/rsearch/energy//pdf/smart-grids_en.pdf European Technology Platform Smart Grids:).

Smart grids are complex organizations of networks regulating, distributing, storing, and generating electrical power. There are amazing analogies in natural and technical networks which should be analyzed by students in interdisciplinary studies. The structure and dynamics of smart grids have surprising similarity with complex protein networks in systems biology regulating the energy supply of a cell. The intelligence of smart grids increases with their ability of self-organizing information processing for optimal energy supply. In communication networks, appropriate prices of optimal energy supply could be automatically negotiated by *virtual agents*. In smart grids, the energy system grows together with information and communication technology in a kind of symbiosis.

6.1.7 Example: Automatic Negotiations of Virtual Agents

A well-known problem with wind mills and solar cells is the unpredictability of production depending on changing weather conditions. In intelligent networks, the need can be locally satisfied by virtual negotiations. A model assumes the following rules and conditions of negotiating virtual agents (Fig. 6.2: Wedde et al. 2007):

1. The need for renewable energy can be satisfied either in a local regional subnet or between subnets. Reserve capacity is used only in exceptional cases.
2. Energy must be adjusted between different voltage levels or different groups of balance on the same level.
3. Producers are also consumers and vice versa.

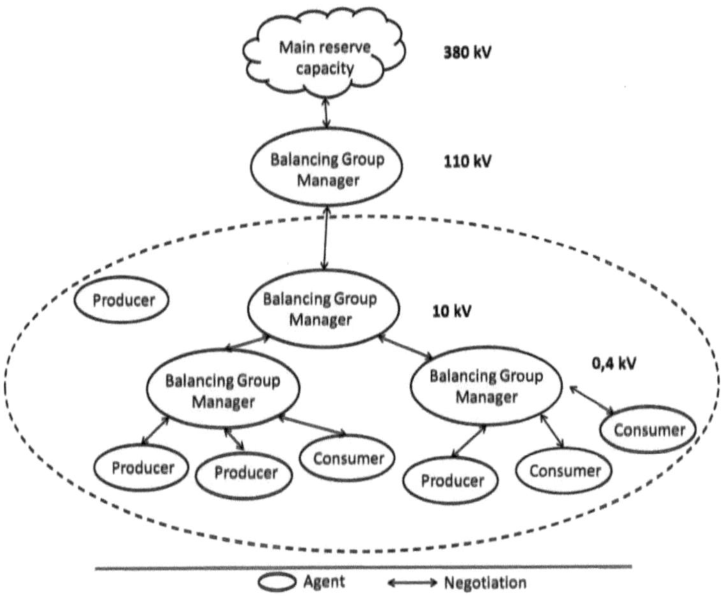

Fig. 6.2 *Smart grid with circles* of balance and assigned agents

4. Negotiations on local energy supply are automatically performed by agents of producers and agents of consumers. They are coordinated by balance group managers working parallel and synchronized in time on each level.
5. In the model, the negotiations start in periods of 0.5 s. The negotiations as well as the distribution of negotiated energy are expected to be finished before the end of each period. Bids and offers arriving in the meantime are negotiated in the next period.
6. At the beginning of each period, each client decides whether he/she takes part as producer or consumer or not. He/she decides with respect to the current difference between the states of demand and production.
7. Bids and offers occur in frameworks of prices with respect to amortization and maintenance. In the model, there are no long-range contracts or discounts for big and future acquisitions which can occur in reality.

The algorithm of negotiation assumes a framework of prices for each level of negotiation. Each balance group manager on each level accomplishes a cycle of coordination of 10 turns. Each turn takes 1 ms. After each turn the balance managers test in parallel whether bids and offers are sufficiently similar. If they are sufficiently similar, a contract between the partners is concluded. A fixed amount is added until the stock or demand is spent. The negotiation strategies of a client are given by an opening bid, an opening offer, and parameters of priority and strategy. After n turns, the unsatisfied agents adapt their bids and offers with respect to an exponential law of behavior which is useful to realize a fast convergence between bids and offers. The negotiated price is the arithmetic mean between similar values. Unsatisfied clients are passed on to the next level of negotiation. On this level, the framework of prices is reduced to a constant relation. The needs and interests of finally unsatisfied clients are satisfied by a central reserve capacity (but with very bad prices).

Short term fluctuations of consumption in the millisecond to minute interval, which are effected by sudden and unpredicted local or regional causes, are not only observed as perturbations in households, but they can endanger the stability of large transport networks. In our model, these critical situations are avoided by the activation of agents after each cycle of negotiation. It is assumed that many electrical appliances (e.g., refrigerator, boiler) can temporarily work without power or with a battery. In these cases, reserve energy can be used for other purposes. The reserve energy is more competitive than the traditional one, because of low costs of transport and storage in the network. Additionally, the balance managers act on each level in parallel in shortest time.

In this way, smart grids with integrated communication systems accomplish a dynamical regulation of energy supply. They are examples of large and complex real-time systems according to the principles of cyber-physical systems (Lee 2008). Traditionally, reserve energy which is used to balance peaks of consumption or voltage drops is stored by large power plants. The model of Fig. 6.2 solves the problem by dynamically reacting strategies of negotiation in proper time. The main problem of changing to renewable energies is the great number of constraints depending on questions of functionality as well a security, reliability, temporary availability, tolerance of failures, and adaptability. Cyber-physical systems with local and bottom-up structures are the best answer to the increasing complexity of supply and communication systems (Cyber-Physical Systems 2008). In a technical co-evolution mankind is growing together with these technical infrastructures in a kind of superorganism.

6.1.8 Example: Internet of Things

Increasing computational power and acceleration of communication need improved consumption of energy, better batteries, miniaturization of appliances, and refinement of display and sensor technology (Weiser 1991; Hansmann 2001). Under these conditions, intelligent functions can be distributed in a complex network with many multimedia terminals. Therefore, the paradigm shift from concentrated computer power in a single computer to distributed computer functions in many functional devices was called *ubiquitous computing*. In a next step, together with satellite technology and global positioning systems (GPS), electronically connected societies are transformed into cyber-physical systems. They are a kind of symbiosis of man, society, and machine. Communication is not only realized between human partners with natural languages, but with the things of our world. Cyber-physical systems also mean a transformation into an Internet of the things. Things in the Internet are equipped with sensors and intelligent computer functions enabling them to communicate with one another.

A commercial application is the concept of industry 4.0: Industry 1.0 was the industry of steam engines. Industry 2.0 was the world of Henry Ford with assembly lines of production, e.g., in car industry. In industry 3.0, industrial robots were additionally applied at an assembly line, replacing human routines by automata. But, industry 4.0 means the internet of things in industry. Work pieces are communicating with their work bench, transport systems, logistics, and marketing, in order to organize their production and sale. On demand production and tailored design according to individual preferences will become the standard procedure in industry 4.0.

6.1.9 Complexity and Big Data Technology

The increasing complexity and globalization of humankind is connected with a huge amount of information and communication networks. There are two main drivers of information and communication technology (ICT): *Moore's law* of increasing computational capacity and *Big Data*. Since many decades, it is verified that, according to Moore's law, the computational capacity of computers is doubled in periods of 18 months with increasing miniaturization and price-reduction of computational devices. In the meantime, the computational speed of supercomputers is measured in petaflops (peta $= 10^{15}$, flop $=$ floating point). Big Data means a giant mass of data which is also measured in peta size. Google, for example, processes 24 petabytes per day, i.e. 6000 times the data stored in the US library. Beyond transactional data stored in relational databases, there are less structured data of weblogs, social media, email, sensors, and photographs that can be mined for useful information. Thus, big data means an amorphous and messy mass of structuralized, less structuralized, and unstructuralized data which cannot be stored and processed by traditional databases and algorithms (Mainzer 2014; Mayer-Schönberger and Cukier 2013).

New algorithms like *MapReduce* (which is used by Google's search engine) or *Hadoop* (programmed in Java) are applied with great success to find correlations and patterns in a data mass. MapReduce uses the functions "*map*" and "*reduce*" which are well-known in functional programming. The function "*map*" separates a messy data mass in several partial packages of data which are mined in parallel computation. The function "*reduce*" integrates the partial results of data mining in a final result. Thus, big data algorithms process all data in a messy data mass, in order to compute correlations and to derive predictions. That is quite different to classical search procedures of statistics using representative samples which are extrapolated for predicting. Further on, a main difference to traditional search procedures is the use of metadata. Metadata allow complete ignorance of meaning and contents of messages and data. In the case of e-mails, only sender and receiver must be known. Signals of mobile phones and GPS-data are also used as metadata, in order to localize the changing positions of persons. In 2009, Google could predict the emergence of an epidemic many weeks before public health officials could react by contacting physicians' offices and extrapolating statistical samples. The Google search machine only computed correlations in changing behavioral patterns of people. Thus, big data technology is not interested in artificial intelligence, but only the application of rapid algorithms to huge data masses, in order to compute probabilities of correlations.

In business and markets, it makes sense to get quick information about probabilistic correlations and tendencies without knowing the causal reasons. Big data algorithms can calculate the probabilistic profile of products and preferences which can be used for improving profit expectation. In economy, big data opens new avenues for chains of economic values. People may earn money by selling or loaning data sets, by offering their skills and know how as data experts, or as entrepreneurs with new business ideas of data application. Big data are the raw oil of the future, but with unlimited possibilities of recycling. In different contexts, data sets can be used for different predictions in different fields of application. But big data is about *what*, not *why*. This is an important point which must be discussed with *students in interdisciplinary courses*:

6.1.9.1 Example: Big Data Correlations in Medicine

Big data technology can be used to make better diagnostic decisions when caring for premature babies. The software processes all patient data in real time in parallel measuring of several quantities in 16 different data streams, such as heart rate, respiration rate, temperature, blood pressure, and blood oxygen level, which together amount to around 1260 data points per second. The system uses parallel computing to detect tiny changes of health states of premature babies. By that, it was possible to predict an infection 24 h before symptoms appear. Actually, at that point of time, very constant vital signs are detected prior to a serious infection. The software tells *what*, not *why*. What big data indicates is a correlation of all data, not causality. Nevertheless: big data saves lives as early warning system.

But, there are clear *limits of big data technology in education* and elsewhere. Big data mining can only detect correlations in huge data sets to predict more or less probable trends. Big data production is not sufficient for good science, business, and governance. But why should we worry about causal laws, when rapid search machines find answers and solutions, before we discover and understand their reasons? In a century of extreme acceleration and progress, only quick reactions and decisions seem to be successful. Therefore, Chris Anderson, former editor-in-chief of the magazine Wired, claimed "*the end of theory*". Big data makes scientific methods obsolete. But his message of the "end of theory" is not only stupid, but even a dangerous technological ideology:

During the last global financial crisis in 2008, many people used financial instruments promising rapid profits without understanding the conditions and foundations of their mathematical foundations. They ended in ruin and catastrophes. In medicine, data correlations without causal explanations deliver no understanding of diseases (e.g., cancer). If you ever have seen the complex molecular interactions in a cell through a microscope, then you get highly respect for the complex mechanisms and laws of life which must be understood for sustainable healing. *In nature, decentralized self-organization and centralized control seem to develop in a well-balanced harmony.*

Big data itself is founded on *theory* in computer science. We must understand the logical and mathematical foundations of algorithms and formal systems, in order to recognize the possibilities, but also the limits of software engineering. Data correlations without philosophy and systems theory deliver no understanding of complex problems. In short: theories without data are obviously *empty*, but big data without systems theory and philosophy is *blind.*

6.2 Reflections

Verena Huber, Klaus Mainzer, and Joachim Schütter

During the workshop the discussion was focused on different aspects: Sustainable infrastructure and supply, sustainable economy, humane living conditions, communication/ mobility/ logistics. The group of experts which was very interested in the aspects of communication, mobility and logistics discussed communication as baseline for analyzing the balance between centralized and decentralized systems.

In this context three kinds of communication were discussed:

- Face to face–communication,
- Internet-communication of people,
- Internet of things (cyber physical systems, industry 4.0).

  In the internet of things the digital world growth together with the physical and societal infrastructure by sensor and robotic technology generating networks of logistics.

How can these different ways of communication be a baseline for describing centralized or decentralized processes? To find balances between centralized and decentralized processes it needs a detailed consideration. And communication has of cause not only a technical but also a social and political dimension.

For a detailed view two examples of production processes were analyzed:

- Car industry
- Food-production (agricultural industry?)

Both production processes are definitely complex systems. And both systems have centralized and decentralized aspects.

Aspects of the balance between centralization and decentralization in food Production:

Farming as basic process of food production should *be organized as a decentralized system* using the decentralized internet technology (Internet-communication is a typical technology of decentralized computer networks.). The benefits of the internet open new avenues of best practice.

There is also a tendency of centralization coming from the EU—Administration of Agriculture. In addition, *information* about the food production process *must be communicated worldwide*—by internet, because the production process needs information like weather, crop-diseases.. ..

Furthermore communication systems enable farmers to use *decentralized information* and signal technology (tractor navigation by GPS). In "agriculture 4.0" it will even be possible to communicate with machines (internet of things) and to produce on demand (tailored food production according to the desires of consumers).

And for the implementation of standards *a centralized approach is needed to implement standards* (in quality, ethics, law, environmental standards. . .).

Aspects of the balance between centralization and decentralization in car production:

A review:

- Industry 1.0 was industrialization with the steam engine (nineteenth century).
- Industry 2.0 means mass-production with a division labor on assembly-lines (Henry Ford).
- Industry 3.0 was mass production supported by industrial robots and tools of automation.
- Industry 4.0 (engl.: Internet of Things) means application of the internet of things to the industrial world.

So car production was a typical example for centralized mass production (industry 2.0 + 3.0). Car production in the future will offer concepts of decentralized production with the industrial internet (4.0). This kind of production opens new possibilities to generate "concepts of mobility"[1)] instead of single cars. Nevertheless centralized standards of quality (security, emission. . .) must be guaranteed.

A big question is how can standards be implemented in order to guarantee trust in a highly industrialized and automated world like the car production? On the one hand state should guarantee the implementation of standards. On the other hand consumers can actively ask for different standard qualities in a decentralized manner. Consumers are better informed by the option of the new media and therefore they demand more participation in political, social and economic decisions. Nevertheless representative elements of democracy are necessary to avoid populism and to support reliable policies.

In individual cases Standards must not always be centralized but can also be introduced by decentralized benchmarking. Industry 4.0 in car production needs trust of the general public. And trust is not only guaranteed by technical security. In spite of all high-technology we may not forget that humans should keep the governance in a high industrialized and automated world. The more human responsibility in a high industrialized system the more trust to the system does people have. And human responsibility must always be defined with respect to different degrees of centralized and decentralized complexity.

References

References of Section 6.1

Berners-Lee T (1999) Weaving the web: the original design and ultimate destiny of the world wide web by the inventor. HarperSanFrancisco, San Francisco

Cyber-Physical Systems (2008) Program announcements & information. The National Science Foundation, Arlington, 30 Sept 2008

Embrechts P, Klüppelberg C, Mikosch T (2003) Modeling extremal events for insurance and finance, 4th edn. Springer, Berlin

Gershenfeld N (1998) The nature of mathematical modeling. Cambridge University Press, Cambridge

Goodwin RM (1990) Chaotic economic dynamics. Clarendon, Oxford

Hansmann U (2001) Pervasive computing handbook. Springer, Berlin

Lee E (2008) Cyber-physical systems: design challenges. Technical report no. UCB/EECS-8. University of California, Berkeley

Lotka AJ (1956) Elements of mathematical biology. Dover, New York (Reprint of the first publication 1924)

Mainzer K (2007a) Der kreative Zufall. Wie das Neue in die Welt kommt. C.H. Beck, München

Mainzer K (2007b) Thinking in complexity. The computational dynamics of matter, mind, and mankind, 5th edn. Springer, New York

Mainzer K (2014) Die Berechnung der Welt. Von der Weltformel zu Big Data. C.H. Beck, München

Mainzer K, Chua LO (2013) Local activity principle. Imperial College Press, London

Mandelbrot BB, Hudson RL (2004) The (mis)behavior of markets. A fractal view of risk, ruin, and reward. Basic Books, New York

Mayer-Schönberger V, Cukier K (2013) Big data. A revolution that will transform how we live, work and think. JohnMurray, London

McCauley JL (2004) Dynamics of markets. Econophysics and finance. Cambridge University Press, Cambridge

Scheinkman JA, Woodford M (1994) Self-organized criticality and economic fluctuations. Am Econ Rev 84(2):417–421

Wedde HJ, Lehnhoff S, Rehtanz C, Krause O (2007) Von eingebetteten Systemen zu Cyber-Physical Systems. Eine neue Forschungsdimension für verteilte eingebettete Realzeitsysteme. Pearl 2008 – Informatik Aktuell. Aktuelle Anwendungen in Technik und Wirtschaft 12

Weidlich W (2002) Sociodynamics. A systematic approach to mathematical modeling in the social sciences. Taylor and Francis, London

Weiser M (1991) The computer for the 21st century. Sci Am 9:66–75

Yang X-S (2008) Mathematical modeling for earth sciences. Dudedin Academic, Edinburgh

Chapter 7
The Wildbad-Kreuth-Declaration

Peter A. Wilderer and Martin Grambow

Loss of stable functioning of major market mechanisms, decay of good governance in many countries of the world, continuation of pollution and excessive exploitation of resources, vanishing interpersonal relationships and last but not least deterioration of aquatic and terrestrial ecosystems encouraged a group of 39 scientists, representatives of regulatory agencies, NGOs, businesses and from media to explore whether shifting from globalization towards decentralization would re-stabilize the Earth system including its human dominated components. The workshop was entitled "Global Stability through Decentralization?—In Search for the right balances between central and decentral solutions". The participants finally resolved the following recommendations:

7.1 Recommendations

§1 **It is an illusion to consider natural ecosystems a normative model for human-made systems. The human capacity for imagining alternative futures and ethical social responsibility distinguish human-made systems calling for unique approaches towards resilience and sustainability.**

§2 **Resilience requires a high level of adaptability. Through maintaining ecological and social diversity a reservoir of options and opportunities**

P.A. Wilderer (✉)
Institute for Earth System Preservation (IESP), Technical University of Munich, Garching, Germany
e-mail: peter.wilderer@mytum.de

M. Grambow
Bayerisches Staatsministerium für, Munich, Germany
e-mail: martin.grambow@stmuv.bayern.de

P.A. Wilderer, M. Grambow (eds.), *Global Stability through Decentralization?*,
Strategies for Sustainability, DOI 10.1007/978-3-319-24358-0_7

are preserved that ensure adaptability. Redundancy buffers systems against shocks and allows to keep them functioning.

§3 **Effective protection of the global commons is a prerequisite for any governance approach, be it central or decentral.**

§4 **Existing governance systems based on the principle of subsidiarity should shift towards decentralized decision making, policy implementation and controlling. In this approach the distribution of power and control lies with the most appropriate agent and all agents are empowered to act according to the best interest of all stakeholders on the basis of fundamental civil rights and obligations.**

§5 **System science with modern communication technologies (cyber physical systems) allow for extensive distributed sensing and control, thereby providing the technology for "distributed intelligence", the prerequisite for future shifts of anthropogenic systems towards more decentralization.**

§6 **Governance systems need democratic legitimacy to assure social support, checks and balances.**

§7 **Positive and negative case studies are effective means to inspire global learning.**

§8 **To avoid depletion of material resources, violation of ecological values and loss of cultural diversity the trend towards globalization must be broken and replaced by new economic approaches (e.g. circular economy) tailored to local conditions.**

§9 **Product functions should gain priority over product ownership.**

§10 **To ensure that anthropogenic systems become and remain sustainable requires a tight interplay of all major societal resources: ecology, economy, science and technology, politics and civil society. Each has to play a specific role in achieving effectiveness, efficiency, social cohesion and resilience, while providing legitimacy to the overall process.**

Signed by:

Klaus Arzet, Friedrich Barth, Werner Bauer, Jürg Bloesch, Carolin Boeker, Asher Brenner, Josef Bugl, Martin Buss, Patrick Dewilde, Jürgen Geist, Martin Grambow, Franz-Theo Gottwald, Bettina Haas, Wolfgang Haber, Sandra Hirche, Verena Huber, Ruixia Liu, Willi Kiefel, Martin Korndoerfer, Amitabh Kundu, Jelena Kurz, Klaus Mainzer, Wolfram Mauser, Andreas Otto, Ulrike Potzel, Ortwin Renn, Verena Risse, Bernhard Schätz, Joachim Schütter, Hugo Spowers, Martin Steger, Bernhard Stoeckle, Katharina Stöckl, Gabi Toepsch, Michael von Hauff, Raoul Weiler, Christian Werthmann, Martin Wilderer, Peter A. Wilderer.

7.2 Explanations

§1 It is an illusion to consider natural ecosystems a normative model for human-made systems. The human capacity for imagining alternative futures and ethical social responsibility distinguish human-made systems calling for unique approaches towards resilience and sustainability.

The species *Homo sapiens* is part of the biosphere on planet Earth and therefore humans, like all other creatures, depend on fundamental natural laws and ecosystem function. Hence, ecosystems should be kept healthy and must not be degraded.

The ecosystem's resilience and function are mainly maintained by the process of self-regulation. In contrast, human-made systems are governed by human traits such as free will, anticipation and eagerness to avoid extinction of humankind at large.

Maintenance of stability of anthropogenic systems is a cognitive process which requires a hierarchical control. Setting and enforcing norms and laws is to be understood as an intrinsic democratic process. To be effective and fair, regional and local conditions must be accounted for as an ethical prerequisite in the pursuit of maintaining and strengthening the state of resilience and stability of regions, and subsequently of the globe as a whole. Implementation implies better insights into the drivers of institutional and individual behavior.

Natural ecosystems evolve around certain environmental opportunities and constraints. In contrast, human systems evolve along a trajectory of awareness and agency. We as human beings, as a community, as a nation, decide where we as a species want to go and which kind of life our environment should enable. Similar to natural ecosystems, the coordination of these nested systems relies on feedback loops, which require honesty in the statement of the intentions and goals set at each level of the nested system.

Achieving resilience and subsequently sustainability does not merely depend on finding technical solutions even if these are inspired by nature itself. Instead, ethical concerns together with religious beliefs and philosophical approaches are likewise important and should not be ignored as they are pertinent in several respects. First, they can help us clarify the relationships between human beings as well as their position in relation to nature and other species. Secondly, they can help individuals to get a grip on their own role, concerns and rights within the process of adaptation.

When choosing ethics as a guiding principle of behavior, a strong distinction between "idealistic ethics" and "realistic ethics" is to be made. The first refer to what a community may consider "the most desirable behavior", while the second refers to what its constituents are actually doing. In order to empower communities to act in accordance to their ideal, confounding conditioning factors and deceptive motivation have to be removed, the most important of which is economical, but also other factors such as ideology, indoctrination, group behavior and peoples quest for pleasure. The community or society has to be open to frequent adjustments, re-negotiations and consensus finding of what the 'most desirable behavior' really is.

§2 Resilience requires a high level of adaptability. Through maintaining ecological and social diversity a reservoir of options and opportunities are preserved that ensure adaptability. Redundancy buffers systems against shocks and allows to keep them functioning.

Changes of climatic, economic, political and technological conditions require simultaneous adaptation. Insufficient system elements need to be improved or replaced by others of higher capacity to cope with novel challenges. A natural ecosystem is in a continuous process of alteration in exchange with other ecosystems (co-evolution). It appears that those shifts can be managed the better the more diverse and redundant components are built into the system.

This concept applies also for anthropogenic systems. For instance, monocultures in agriculture as well as monopolies in the economic arena are known to be vulnerable to disturbances whereas diverse systems have a better chance to remain in a dynamic equilibrium. Robustness in the face of varying local conditions can be readily achieved in systems with a high degree of diversity and redundancy. Adaptive properties are based on overall conditioning of the system.

§3 Effective protection of the global commons is a prerequisite for any governance approach, be it central or decentral.

Global commons refer to the use and ownership of vital resources shared by humanity at large. The World Conservation Strategy proposed by the International Union for Conservation of Nature and Natural Resources (IUCN) jointly with UNESCO, the United Nations Environment Programme (UNEP) and the World Wildlife Fund (WWF) considers the Earth's surface beyond national jurisdiction as generic global commons. Included in the list of global commons are the open ocean, the atmosphere, the Polar Regions, but also the outer space and the cyberspace. This encompasses an intact environment, i.e. well functioning ecosystems as the basis for sustainable use of these global commons.

As outlined in the Wikipedia chapter on Global Commons, their management requires pluralistic legal entities, usually international and supranational, public and private, structured to match the diversity of interests and the type of resource, and stringent enough with adequate incentives to ensure compliance. Such management systems are necessary to avoid, at the global level, the classic tragedy of the commons, in which common resources become overexploited or ecosystems are abused as waste disposals.

The authors of this declaration suggest an important upgrading of the list of global commons to resources such as safe drinking water and fertile soil. Access to electricity, medical care and reasonable income should also be considered as global commons. The community of States as well as national and local communities are encouraged to take all possible measures to ensure safeguarding such commons in the interest of maintaining and keeping social and economic systems at all levels in the state of long-term stability.

§4 Existing governance systems based on the principle of subsidiarity should shift towards decentralized decision making, policy implementation and controlling. In this approach the distribution of power and control lies with the most appropriate agent and all agents are empowered to act according to the best interest of all stakeholders on the basis of fundamental civil rights and obligations.

Subsidiarity is an organizing principle of decentralization stating that a matter ought to be handled by the smallest, lowest, or the least centralized authority capable of addressing that matter effectively. The concept is applicable in the fields of government, political science, neuropsychology, cybernetics, management and in military command (http://en.wikipedia.org/wiki/Subsidiarity).

However, the principle of subsidiarity is only one way of deciding on the division of power and responsibility between hierarchical levels. There are at least two problems: (1) it is based on hierarchical thinking, and (2) it is rigid and often leads to conflicts between levels. In politics, this can be illustrated by a federalist state organization, where e.g. provincial governments compete with state governments.

A better principle is that of the most appropriate agent, e.g. the one which has to exert the least possible effort to achieve the desired result, or else the agent that has the best expertise. A well working distributed system will most likely be based on a network between largely independent but highly connected agents (a very flat hierarchy), in which power and responsibility is given to the most effective agent, and the network is set up in such a way that empowerment is decentralized to achieve the intended results most effectively.

A good example is found in the modern "Smart Grid", in which every agent (the network authority, energy prosumers, large energy producers, specialized storage agents, users) all have their own power and responsibilities logically fitting together. In that system technological expertise is unevenly distributed but empowerment is evenly distributed. Local agents can make decisions in the best interest of the stakeholders, while network stability is ensured by the network provider. This induces economic incentives with the prosumers ensuring stability. However, every agent in a large distributed network has to be accountable to all stakeholders so that no single agent or group of agents misuses the trust invested in them. Subsidiarity reaches its limits however, if local actions have major repercussions on larger units or are in contrast to fundamental human rights and obligations.

§5 System science with modern communication technologies (cyber physical systems, CPS) allow for extensive distributed sensing and control, thereby providing the technology for "distributed intelligence", the prerequisite for future shifts of anthropogenic systems towards more decentralization.

In a technical evolution, a global communication network (World Wide Web) is emerging with surprising similarity to self-organizing ("decentralized") neural networks provided by evolutionary biology. The motivation for this is clearly that modern communication technologies "break the prisoner's dilemma" by allowing the sharing of strategies, the setting of common beneficial goals and early adaptation to changing conditions. However, this will only function when proper alignment based on a common understanding on benefits is achieved, and distributed conditioning measures are put in place that motivate agents with independent control to make beneficial choices.

Modern information and communication technologies have provided concepts and solutions facilitating a balance between centralized and decentralized systems. Those concepts, e.g., virtualization of resources, self-organization of processes, and hierarchisation of services have demonstrated the ability to substantially increase robustness and adaptability of systems, as seen in large-scale robust internet-based functionalities like VoIP (Voice over IP).

By using the so-called 'embedded systems' technology, i.e. digital information processing combined with analog electric/electronic control, as a bridging technology, these concepts are increasingly incorporated in large-scale systems governing physical (i.e., mechanical, chemical, electrical, etc.) processes as well as 'cyber' (i.e., organizational, economical, etc) processes. These 'cyber-physical systems' like smart traffic systems, smart factories, or smart electric grids, generally use 'fractal' (i.e., hierarchically self-similar) forms of cooperation and coordination, thus achieving the necessary balance between centralized and decentralized governing schemes.

The balance between centralized and decentralized governing schemes enables CPS to provide the core capabilities of cross-X (like cross-organization, cross-domain, cross-discipline), life-X (e.g., life-update, life-reconfiguration, life-extension), and self-X (e.g., self-documenting, self-monitoring, self-healing). Using these capabilities CPS support the construction of architectures and processes robust against and adaptive to unexpected or changing behavior of their users and environments. The practicably of this approach is currently demonstrated by the increasing number of CPS being constructed. For example, in the smart grid domain, the above-mentioned capabilities allow the stable integration of low-volume renewable resource like private-home photovoltaics using coordinated decentralized buffering and load-shifting schemes, implementing the 'cellular' approach for grids and markets asked for by the German Bundesnetzagentur in its 2011 Position Statement 'Smart Grid und Smart Market'.

In the past the classical IT-world was only virtual and separated from physical infrastructures. In CPS, the IT-world grows together with the physical infrastructure of our civilization like the nervous system with an organism. CPS observe their environment by sensors, process their information and influence their environment with actuators according to communication devices. CPS are complex systems of many self-organizing net components, dramatically increasing the adaptability, autonomy, reliability and usability of automotive, aerospace, energy, healthcare, manufacturing, transportation, and consumer appliances—a challenge of human control and responsibility.

In general CPS lead to the next, the 4th industrial revolution. The 1st industrial revolution introduced the steam engine. The 2nd industrial revolution meant centralized mass production, division of labor, and working on the assembly line. The 3rd industrial revolution additionally applied industrial robots for further automation of production. The 4th industrial revolution changes production on the basis of CPS and the "Internet of Things". Production, marketing, and trade are transformed into self-monitoring and self-organizing complex system. Cloud manufacturing connects the "Internet of Things" with cloud computing, supported by VR (virtual reality)-technologies, parallel and distributed working computer nets. Cloud manufacturing leads to decentralized production and trade nets. The working world is organizing itself, supports flexible work of employees, on demand, individual („tailored") service of clients. In contrast, centralized and standard mass production was typical for industry 2.0 and 3.0.

CPS produce a huge amount of data in all domains of science, economy, and society. Big data technology and computing networks open new avenues of fast data mining and profiling of products and persons in economy and society, but also of centralized and totalitarian control worldwide. Contrary to this dangerous misuse, fast algorithms and computing networks should improve human well-being with more secure and efficient, but less vulnerable human infrastructure. Digital dignity is the primary ethical goal in the complex world of Big Data and cyber-physical systems. In the age of globalization, mankind is in an unstable ("chaotic") phase transition of high complexity, depending on innovations of science and technology, risks of ecology, economy and finance, creative chance and innovative change. The nonlinear dynamics of CPS need complexity policies of global governance and controlled emergence to support a balance of centralized and decentralized order.

§6 Governance systems need democratic legitimacy to assure social support, checks and balances.

Democratic legitimacy does not guarantee any of the desirable properties of the Earth system. It is certainly necessary for reasons of societal health, but it has to be flanked with purpose, just as good health does not guarantee good behavior; the issue is "what is good behavior and how can a person be conditioned to behave well?" Much more is needed than democratic legitimacy.

There must be a democratically accepted common direction which might be termed "common ethics". How can this be achieved? Agents act at the various levels of a society. There are several stratifications, not only government strata. To be considered are also functional strata, such as business systems, interrelated financial institutions, service sectors including learning, knowledge and information agents.

What is needed is an "alignment of intelligence" between all these intelligent institutions. The societal system has to provide the means for such an alignment. The latter is a kind of "democratic legitimacy", but not one in the usual sense as majority agreement. Understanding how it comes about and how it can be propagated and adopted is essential for our goal to achieve a sustainable earth system.

The more centralized governance systems are, the more it is essential to accomplish a well-rounded balance between effectiveness, efficiency, resilience and social cohesion. Effectiveness refers to the need of societies to have a certain degree of confidence that human activities and actions will actually result in the consequences that the actors intended when performing them. Efficiency describes the degree to which scarce resources are used to reach the intended goal. The more resources are invested to reach a given objective, the less efficient the activity under question remains. Resilience describes the capacity to sustain functionality of a system or a service even under severe stress or unfamiliar conditions. Finally, social cohesion covers the need for social integration and collective identity despite plural values and lifestyles.

All four needs or functions of society build the foundation for legitimacy. Legitimacy is a composite term that denotes, first, the normative right of a decision-making body to impose a decision even on those who were not part of the decision-making process (issuing collectively binding decisions), and second, the factual acceptance of this right by those who might be affected by the decision. These two conditions of legitimacy can best be accomplished by assuring a transparent and inclusive process of decision making (social support) and the implementation of an effective controlling process by independent agencies (such as the court of justice) as a means to evaluate the consequences of political interventions and review these consequences on the four key criteria (checks and balances).

§7 Positive and negative case studies are effective means to inspire global learning.

It is well known that we learn best from failures (negative case studies). It is also well known that success (positive case studies) triggers excitement and motivation—most important preconditions in the process of finding and implementing solutions of burning problems. Obviously, we need both positive and negative experience to drive learning processes, to get in the position to create innovative thinking and thus respond proactively to changes of the conditions we get confronted with.

However, negative outcomes of case studies are often associated with the presumption the persons in charge (scientists as well as entrepreneurs and politicians) would be unable to perform properly. Subsequently, negative experiences are rarely presented as an opportunity to learn from. Worse than this, negative results tend to make the public concerned, often bewildered, even paralyzed. Press media use this effect to enhance audience rates and sales figures. Consternation, however, does not contribute to the solution of problems and to progress in learning. In a time of rapid global changes it appears of upmost importance to take any possible attempt to raise awareness of the importance of case studies that should not be a priori expected to deliver positive results only. It is the responsibility of academia to convey the knowledge that enables decision makers as well as the general public to draw the right conclusions from positive and negative cases. Only then humanity gets in the position to overcome global as well as local threats.

A crucial measure to minimize or avoid fundamental errors and mistakes that are based on ignorance is education. Therefore, countries are encouraged to establish sound and efficient education systems for teachers, children and adults (including the use of internet and new ICT) to increase ecological and social competence.

§8 To avoid depletion of material resources, violation of ecological values and loss of cultural diversity the trend towards globalization must be broken and replaced by new economic approaches (e.g. circular economy) tailored to local conditions.

Since economy is the key driver for human activities, a paradigm change needs to be initiated and directed towards sufficiency. Quantitative growth (e.g., human population) and qualitative growth (e.g., wealth) are subject to natural laws and cannot be unlimited. Technical efficiency cannot compensate excessive growth. Our ecological footprint needs to be significantly reduced. Non-monetized natural values (ecosystem services) must be defended against economic pressures as they are needed for human well being. Some principles such as polluter/causer pay, precaution, solidarity, fight cause instead of effect, recycling, and public participation should be applied. We need to work with, and not against nature.

Faced with increasing resource prices and dwindling reserves, different economical approaches have already been suggested. Among the most discussed are the bio-economy, the blue economy, the circular economy, decentralized water management, zero waste economy, the economy for the common good and the factor-10-economy. However, these approaches should be moved from the exclusive academic/political/industrial arenas. They need to be properly discussed in and by the public with the aim of reaching a broad consensus. This step should include local, national or global networking in the fields of knowledge, ecology, technology and finance, representing real decentralization. At the same time, the consensus reached will be democratically legitimized by design, which will facilitate the implementation of a novel economic system by political leaders.

The issue here is "how to get an optimal division of responsibilities, given an agreed standard of optimality?" Arrangements have to be made at the various levels of responsibility (up to the global level) in such a way that each participating agency sees it as its advantage to work towards that optimal solution which actually may change and evolve over time. There are two sides to this issue: at each level, control measures have to be appropriate for that level, and benefits and profits have to be shared between all participating agents (including the consumers), in proportion to their efforts. One cannot avoid making intelligent blueprints that can be understood and underwritten by all concerned needing honest thinking at all levels, so it is a primordial ethical issue.

Modern water management can serve as such a blueprint, as these new concepts and strategies are integrative, transdisciplinary, often transboundary and complex, encompassing whole river basins. They need to be properly designed, operated and maintained.

§9 Product functions should gain priority over product ownership.

The value of functionality over ownership is to be favored in both, a centralized or decentralized environment. Selling the service that reliably functioning products provide over adequate and clearly communicated defined use periods (performance leasing e.g., mobility of a car, manufacturing robots, mobile phones) instead of turning over the ownership for the product materials, will incentivize cleaner production, use of better materials and improving holistic product quality.

By keeping the ownership over materials, the current incentive for increased material turnover rates will be replaced by a strong incentive to become a material steward aiming at the return of every ounce of material and re-using the resources in new products. In combination with decentralized service, repair and upgrading enhancement of the advantage of regions and local communities (e.g., workplaces, income, tax revenues etc.) is to be expected.

Sometimes concern is voiced that product owners take more care of it than mere product users. However, this could be overcome by applying the highest possible fraction of recyclable products where the recycling process is also driven by renewable energy generating materials, as well as the return logistics. Another approach could be to implement incentives that reward longer service contract. In this model the consumer would pay less and less per month, while continuation of driving the leased car, and using laptop, mobile phone etc. is secured.

Ownership of things is traditionally valued as an expression of independency in using and caring for goods once purchased. Shift from ownership to leasing bears the risk of losing liberty while becoming overly dependent on service providers. Those concerns must be treated with absolute carefulness. Strengthening competition between product producers (soft- as well as hard-ware) and service providers appears to be an important control mechanism.

§10 To ensure that anthropogenic systems become and remain sustainable requires a tight interplay of all major societal resources: ecology, economy, science and technology, politics and civil society. Each has to play a specific role in achieving effectiveness, efficiency, social cohesion and resilience, while providing legitimacy to the overall process.

Centralization is driven by the role capital plays in our societies—capital not only in monetary terms but also in terms of land and physical resources. Ever since the globalized bank system has been largely decoupled from real-world assets or even real economic growth, the physical limitations for centralization do no longer exist. Hence, to achieve a balance between centralization and decentralization it is now high-time for re-orientation towards decentralized solutions in all sectors of our economy.

As mentioned in §1 free will, but moreover impatience and greed dictate the general behavior of humans. This often leads to corruption and illegal connections, since money and wealth provide power. The embedded self-regulation capacity of economic and societal systems is often hampered by the lack of empowerment and solid knowledge at the local level. Control by central authorities therefore seems as inevitable as control by an independent law authority. The alignment and empowerment of independent and potentially hostile agents may be facilitated through advanced network technology. This is a field that needs considerable further development. While decentralization, besides responding to local concerns in a more effective manner, would be a powerful tool to get the local voices and concerns heard at the national and global level, it is difficult to ensure equity without a framework for guiding the future developments of science and technology at national and global level. Thus, there is a strong case for designing an institutional and legal system for guiding the future developments in this area.

The foundation of sustainable development is the need for a well-rounded balance between effectiveness, efficiency, resilience and social cohesion as explained in §6. Within the macro-organization of modern societies, these four functions are predominantly handled by different societal systems: economy, science (expertise), politics (including legal systems), and the social sphere. Another way to phrase these differences is by distinguishing among competition (market system), hierarchy (political system), and cooperation (socio-cultural system). These insights suggest that for complex policy decisions that are crafted to enhance the sustainability of society, representatives of all four sectors of society need to be included in order to ensure that decisions are effective, efficient, resilient and fair.

Network technology, similar to ecosystem networks, has the great potential to combine efficiently decentralized local concerns and demands with centralized national and international interests. Thus, guiding the future developments of science and technology is required, together with a global framework of an institutional and legal system to provide more equity between nations.

Erratum to: Global Stability through Decentralization?

Peter A. Wilderer and Martin Grambow

P.A. Wilderer, M. Grambow (eds.), *Global Stability through Decentralization?*,
Strategies for Sustainability, DOI 10.1007/978-3-319-24358-0_5

Corrections to chapter 5

The original chapter contains incorrect Figures 5.1, 5.3 and 5.4. We have corrected the figures in the chapter as mentioned below.

Chapter 5.1: page 141; Figure 5.1 and 5.4: The drawing in Fig. 5.1 has been replaced by the drawing presented in Figure 5.4, and vice versa.

Chapter 5.1: page 143; Figure 5.3: The drawing has been replaced by the drawing in the next page.

The online version of the original chapter can be found at
http://dx.doi.org/10.1007/978-3-319-24358-0_5

P.A. Wilderer (✉); Corresponding author for Section 5.1
Institute for Earth System Preservation (IESP), Technical University of Munich, Garching, Germany
e-mail: peter.wilderer@mytum.de

M. Grambow (✉); Corresponding author for Section 5.1
Bavarian State Ministry of the Environment and Consumer Protection, Munich, Germany
e-mail: martin.grambow@stmuv.bayern.de

P.A. Wilderer, M. Grambow (eds.), *Global Stability through Decentralization?*,
Strategies for Sustainability, DOI 10.1007/978-3-319-24358-0_8

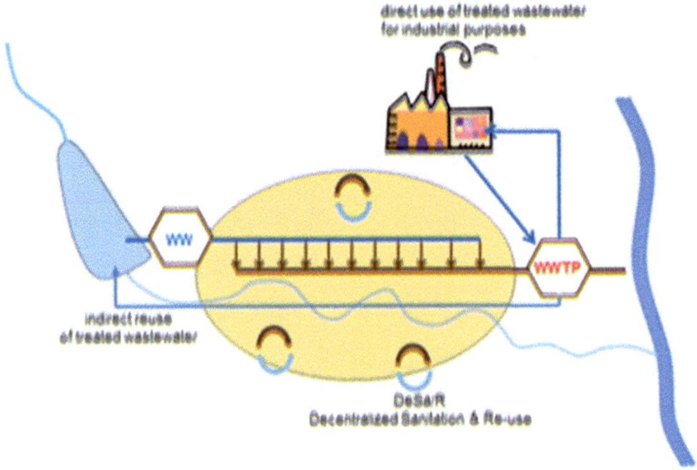

Fig. 5.3 A closer look to options for cascading use of water, and for direct and indirect water re-use in urban areas

Appendix

Zugspitze Declaration on the Responsibility of Humanity for the Functioning of the Earth System

Faced by serious challenges to the Earth system, a group of scientists, politicians and business representatives met for four days in Wildbad Kreuth, Germany, at a workshop on "**Earth-System Engineering: The Art of Dealing Wisely with the Planet Earth**". The participants agreed to the following declaration:

§1 Our global crisis

Global climate change threatens the life-enabling functions of the Earth system.

§2 We ought to learn from the past

We should be able to tackle climate change as effectively as the problems resulting from acid rain and stratospheric ozone depletion were mitigated. But we will need much more concerted action.

§3 Our deeds must follow our words

There is plenty of knowledge available to handle this global crisis. Let us get started!

§4 We must raise awareness of the problem

Behavioral changes will be necessary to solve the Earth system crisis.

§5 We need to protect our indispensible global commons

To regulate usage of the global commons, including the atmosphere and oceans, an enforceable international legal framework needs to be formulated, adopted and applied globally.

P.A. Wilderer, M. Grambow (eds.), *Global Stability through Decentralization?*,
Strategies for Sustainability, DOI 10.1007/978-3-319-24358-0

§6 Execution of deliberate climate modification requires authorization

Climate engineering should only be permitted after rigorous assessment and authorization by a process of international consensus to which all nations are supposed to contribute.

§7 Sustainability has to be our goal

Our global society must understand that human endeavors can only be tolerated when sustainability criteria are met.

§8 A global commons trust fund should assist in financing advanced research

A global commons trust should be made available to finance advanced research which is holistic and multi-disciplinary in nature.

§9 We need a United Nations authority responsible for preserving the functioning of the Earth system

A high-level United Nations body is to be established and authorized to enforce measures capable of preserving the life-enabling functions of the Earth system.

This declaration was signed by:
Dr. Padam Bhojvaid, Dr. Josef Bugl, Dr. Paul Josef Crutzen, Elena Davydova, Dr. Helmut Fluhrer, Dr. Martin Grambow, Dr. Michael von Hauff, Dr. Rafaela Hillerbrand, Dr. Eva Lang, Dr. Tim Lenton, Dr. Hamish McGowan, Dr. Wei Meng, Dr. Lee Miller, Dr. Rolf Müller, Dr. Deb Niemeier, Fred Pearce, Dr. Ulrike Potzel, Yuri Saveliev, Dr. Yong Hui Song, Dr. Akimasa Sumi, Dr. Naomi Vanghan, Dr. Raoul Weiler, Dr. Peter A. Wilderer, Dr. Stefan Wuertz

Explanations

§1 Our global crisis
Until recently, the biggest challenges to the survival of the increasing world population were the food and water deficits, pollution, loss of biodiversity, poverty, and inequality. Now, global climate change threatens to further exacerbate all these problems and compromise our efforts to deal with them.

§2 We ought to learn from the past
The world has recently shown the capacity and flexibility to quickly respond to short-term natural disasters. Moreover, environmental management has successfully addressed problems resulting from acid rain and stratospheric ozone depletion. The signatories of the declaration urge that similar capabilities be quickly mobilized to address the far more serious and longer term Earth-system crisis.

§3 Our deeds must follow our words
There is no excuse for delaying implementation of available state-of-the-art technologies. While there is need for innovation in certain areas, proven, established and sustainable responses already exist for handling large parts of the global crisis. The signatories of the workshop want to motivate those who have solutions, in order to share this knowledge with the global community so that implementation can begin now.

§4 We must raise awareness of the problem
Efforts to address the crisis require social empowerment to change behavior and deal wisely with the Earth system. This is the primary responsibility of the current generation.

§5 We need to protect our indispensable global commons
Utilization of the global commons, including the atmosphere and oceans, is currently only partially regulated, and degrading global commons carries few penalties. We recommend the formation of a binding international legal framework and a Global Commons Trust as proposed by Christopher D. Stone (in "Wege zum ökologischen Rechtsstaat" (Path towards an ecological state under the rule of law), H. Baumeister (ed), E. Bottner Verlag, 1993). Most urgent is a globally

P.A. Wilderer, M. Grambow (eds.), *Global Stability through Decentralization?*,
Strategies for Sustainability, DOI 10.1007/978-3-319-24358-0

validated administration of a global cap and redistribution system for atmospheric carbon emission or climate protection credits.

§6 Execution of deliberate climate modification requires authorization
Climate engineering is the deliberate, planned, large-scale intervention in the climate system. Such initiatives with potential global effects need rigorous risk and legal assessment, and authorization by a process of international consensus to which all nations are supposed to participate. Climate engineering must never be seen as an alternative to the necessity to solve the global crisis at its roots.

§7 Sustainability has to be our goal
The purpose of the new frontier of Earth system engineering must be to promote improved management of the global commons. Our global society must find a relationship between the natural environment and human endeavors that ensures sustainability.

§8 The global commons trust fund should assist in financing advanced research
The global commons trust fund should support, among other important tasks, advanced research which is holistic and multi-disciplinary in nature.

§9 We need a United Nations authority responsible for preserving the functioning of the Earth system
A United Nations body is to be established and authorized to enforce measures capable of preserving the life-enabling functions of the Earth system. Decisions should be based on the advice of the IPCC, and on the recommendations of a second intergovernmental panel responsible for assessing and evaluating, within the context of sustainability, those technologies and methods proposed to preserve the life-enabling functions of the Earth system. This proposal goes beyond earlier suggestions to elevate the UN Environment Programme (UNEP) to the level of the WTO, because of the severity of the crises we now face.

Acknowledgement

The authors acknowledge the idealistic and financial support provided by the Bavarian State Ministry for Environment and Consumer Safety. Moreover, the authors are grateful for the support received from the Institute for Advanced Study of the Technische Universität München.

P.A. Wilderer, M. Grambow (eds.), *Global Stability through Decentralization?*,
Strategies for Sustainability, DOI 10.1007/978-3-319-24358-0

Index

A

Adaptive cycles, 64–68, 70
Agriculture, 7, 9, 13, 126–129, 140, 142–145, 147, 148, 160, 179, 184

B

Big data, 165–178, 187
Bioeconomics, 127, 129
Biosphere, 8, 11, 13, 84–90, 95, 96, 98, 102, 104, 127, 183
Biotic regulation, 88–95, 102

C

Centralization, 11, 17, 32, 47, 75–80, 82, 103, 127, 128, 131, 157, 179, 190
Chaos, 24, 25, 27, 28, 30, 31, 84, 87
City planning, 77
Civilization, 83–104, 111, 146, 166, 172, 186
Communication technologies, 60, 166, 182, 185, 186
Consumer, 2–6, 9, 10, 13, 21, 39, 40, 101, 129, 132, 134, 135, 143, 153, 160, 173, 174, 179, 186, 189, 190
Cultural evolution, 6, 8
Culture, 8, 49, 82, 98, 101, 107–112, 126, 153
Cyber physical systems, 17, 60, 63, 64, 172, 175, 178, 182, 186

D

Decentralization, 11, 17, 43, 47, 55–64, 76, 77, 84, 91–93, 103–107, 118, 119, 124–130, 136, 144, 149–152, 165, 179, 181, 182, 185, 189–191
Decision making, 31, 34, 37, 38, 77, 81–83, 110, 132, 141, 146, 148, 182, 185, 188
Duality of nature, 16
Dynamic systems, 12

E

Economic and societal crisis, 1, 190
Economy, 11–16, 66, 81, 102, 109, 112, 117–132, 134, 146, 161, 162, 168–170, 177, 178, 182, 187, 189–191
Ecosystem function and service, 1–10, 111, 135, 136, 140, 183
Ecosystems, 1–17, 46, 64, 66, 69, 83–104, 109–111, 120, 121, 135, 136, 140, 142, 145, 146, 148, 181, 183, 184, 189, 191
Education, 16–17, 33, 38, 107–110, 126, 127, 134, 135, 144, 166, 167, 169, 171, 172, 177, 189
Electrical energy, 32, 38
Energiewende, 35
Energy, 2, 3, 5, 6, 12, 14, 16, 22, 25, 26, 32, 33, 38, 40, 55–64, 67–71, 78, 79, 84–88, 93, 95, 96, 101–103, 122–124, 126–129, 133–135, 144, 147, 148, 155–161, 165, 166, 172–175, 185, 186, 190
Energy crisis, 146, 148
Environment, 1–4, 6–9, 11–14, 17, 28, 31, 34, 36, 37, 43, 46–48, 51, 52, 54, 63, 76, 80, 84, 85, 87–90, 92–95, 97–99, 101–112, 118–124, 126–134, 145, 148, 150–162, 179, 183, 184, 186, 190
Ethics, 11, 16–17, 22, 34–38, 41, 43, 110, 136, 172, 179, 183, 187

P.A. Wilderer, M. Grambow (eds.), *Global Stability through Decentralization?*, Strategies for Sustainability, DOI 10.1007/978-3-319-24358-0

G
Global crisis, 14, 102
Global warming, 112, 135, 141, 148
Governance, 75–112, 130, 132, 133, 136, 152, 165, 166, 177, 180–182, 184, 185, 187, 188

H
Health risks, 149
Human dimension, 7
Human impacts, 14–16
Hydrosphere, 3

I
Industrial parks, 122–123
Informal cities, 77
Information Technologies. *See* Communication technologies
Internet of Things, 84, 101, 175–176, 178, 179, 187

L
Land use management, 146
Limited growth and Technology, 121

M
Man-made systems, 9, 13, 21–71, 181, 183
Micro-pollutants, 151

N
Neoliberalism, 14

O
Organic farming, 17, 129

P
Paradigm of economy, 14
Prosumer, 62, 63, 185

R
Regulation, 1–17, 26, 46–50, 52, 53, 66, 70, 88–95, 102, 103, 134, 136, 166, 171, 175, 183, 190
Renewable energy, 67–70, 102, 112, 124, 133, 173, 190
Resilience, 12, 14, 32, 36, 42, 55, 66, 69, 80–83, 85, 132, 146, 150, 181–184, 191, 188190
Risks, 7, 66, 110, 111, 124, 126, 129–132, 144, 148, 149, 166, 167, 170, 171, 187, 190

S
Sanitation, 46, 47, 51, 77, 139, 144, 145
Smart grids, 26, 32, 38–43, 58, 60–64, 166, 172–175, 185, 186
Sociotechnical systems, 165, 172
Solid waste management, 152–162
Stability, 12, 21, 23, 25, 32, 34–38, 40–42, 55, 59, 64, 84, 90, 92, 95, 98, 101, 132, 135, 136, 142, 146–148, 171, 175, 181, 183, 185
Subsidiarity principle, 50, 103
Sustainability, 13, 14, 16, 25, 32, 34, 36, 42, 64, 82, 83, 85, 93, 109–112, 118, 120–123, 125, 128–132, 135, 140, 147, 149, 150, 153, 181, 183, 191
Sustainable economy, 117–125, 130–136, 178
Sustainable trade, 129
Systems science, 165–167, 169

T
Technology, 11–13, 16, 17, 35, 37, 38, 42, 56, 60, 61, 65, 68, 69, 80, 101, 102, 104, 107, 122, 124, 126, 128, 129, 131, 141, 142, 144, 145, 147–149, 154, 165, 172, 173, 175–179, 182, 185–187, 189–191
Toxic materials, 151
Trade, 13, 109, 126, 129–131, 133–135, 153, 168, 187
Transportation, 22, 24, 25, 32, 76, 100, 134, 139, 142, 143, 156, 186

U
Urbanization, 76–80, 149, 150, 165

W
Waste management, 127, 150, 152–162
Wastewater management, 149–150, 152
Water and food crisis, 148
Water reuse, 139, 143

MIX
Papier aus verantwortungsvollen Quellen
Paper from responsible sources
FSC® C105338

If you have any concerns about our products,
you can contact us on
ProductSafety@springernature.com

In case Publisher is established outside the EU,
the EU authorized representative is:
Springer Nature Customer Service Center GmbH
Europaplatz 3, 69115 Heidelberg, Germany

Printed by Libri Plureos GmbH
in Hamburg, Germany